Praise for Neil deGrasse Tyson

"[Tyson] tackles a great range of subjects . . . with great humor, humility, and—most important—humanity." —*Entertainment Weekly*

"[A] looming figure. . . . [A]n astronomer to the bone."
—Carl Zimmer, *Playboy*

"It's one thing to be a lauded astrophysicist. It's another to possess a gift for comedic timing. You don't normally get both, but that's Neil."
—Jon Stewart, *The Daily Show*

"It's hard to imagine a better man to reboot the cosmos than Neil deGrasse Tyson." —Dennis Overbye, *New York Times*

"[Tyson] is bursting with ideas." —Lisa de Moraes, *Washington Post*

"It's more imperative than ever that we find writers who can explain not only what we're discovering, but how we're discovering it. Neil deGrasse Tyson is one of those writers." —Anthony Doerr, *Boston Sunday Globe*

"Tyson . . . is a confidently smooth popularizer of science." —*People*

"The heir-apparent to Carl Sagan's rare combination of wisdom and communicative powers." —Seth MacFarlane, creator of *Family Guy*

SPACE CHRONICLES

ALSO BY NEIL DeGRASSE TYSON

The Pluto Files: The Rise and Fall of America's Favorite Planet

Death by Black Hole: And Other Cosmic Quandaries

Origins: Fourteen Billion Years of Cosmic Evolution

FACING THE ULTIMATE FRONTIER

W. W. Norton & Company • New York • London

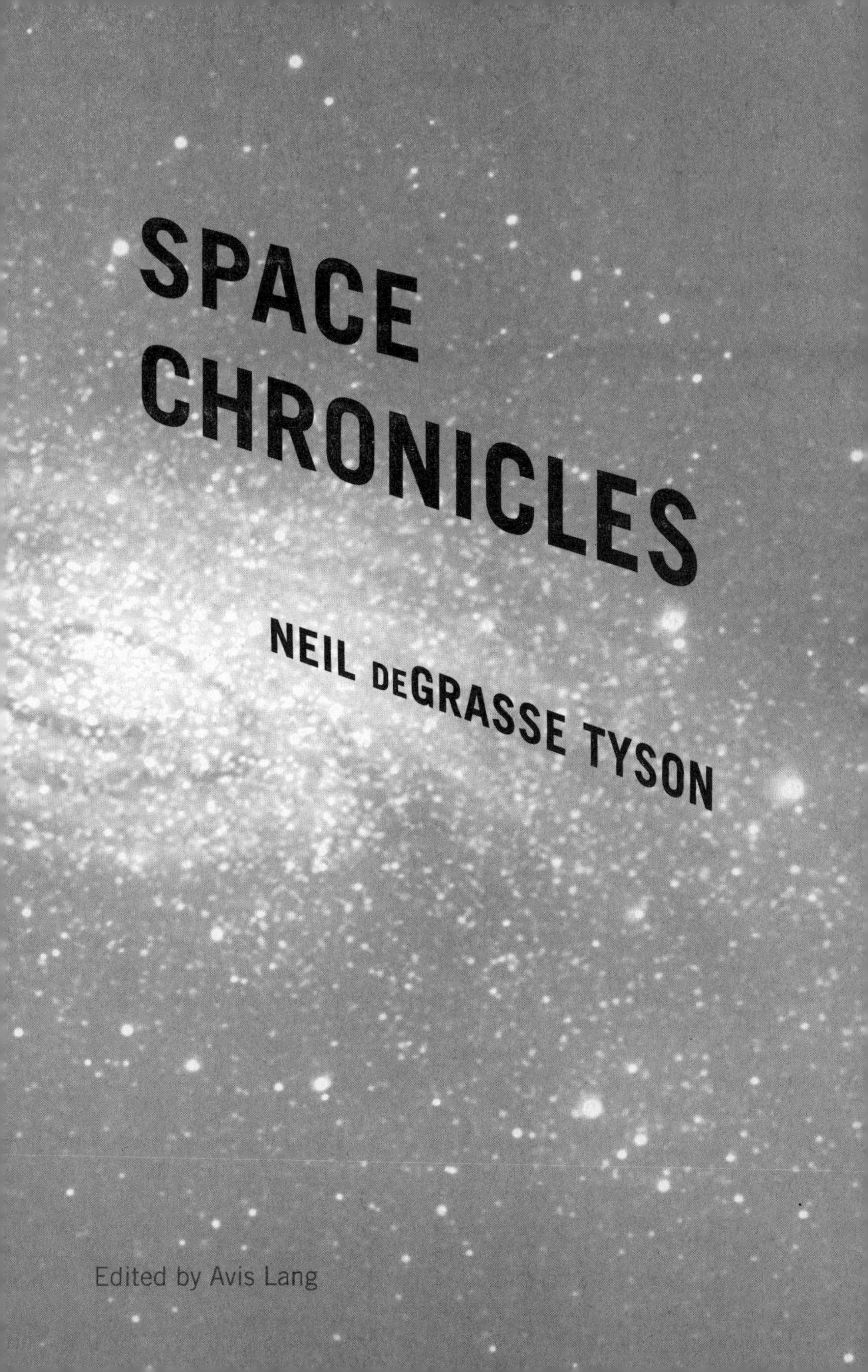

SPACE CHRONICLES

NEIL deGRASSE TYSON

Edited by Avis Lang

Printed in the United States of America
First published as a Norton paperback 2013, reissued 2014

For information about permission to reproduce selections from this book, write to Permissions, W. W. Norton & Company, Inc., 500 Fifth Avenue, New York, NY 10110

For information about special discounts for bulk purchases, please contact W. W. Norton Special Sales at specialsales@wwnorton.com or 800-233-4830

Manufacturing by RR Donnelley, Harrisonburg
Book design by Judith Stagnitto Abbate / www.abbatedesign.com
Production manager: Devon Zahn

Library of Congress has cataloged the hardcover edition as follows:

Tyson, Neil deGrasse.
Space chronicles : facing the ultimate frontier / Neil deGrasse Tyson ; edited by Avis Lang. — 1st ed.
p. cm.
Includes index.
ISBN 978-0-393-08210-4 (hardcover)
1. Astronautics and state—United States. 2. United States. National Aeronautics and Space Administration. 3. Manned space flight—Forecasting. 4. Outer space—Exploration. I. Lang, Avis. II. Title.
TL789.8.U5T97 2012
629.40973—dc23

2011032481

ISBN 978-0-393-35037-1 pbk.

W. W. Norton & Company, Inc.
500 Fifth Avenue, New York, N.Y. 10110
www.wwnorton.com

W. W. Norton & Company Ltd.
Castle House, 75/76 Wells Street, London W1T 3QT

1 2 3 4 5 6 7 8 9 0

To all those who have not forgotten how
to dream about tomorrow

CONTENTS

EDITOR'S NOTE

Back in the mid-1990s, Neil deGrasse Tyson began writing his much-loved "Universe" column for *Natural History* magazine. At that time, the magazine was hosted, both financially and physically, by the American Museum of Natural History, which also hosts the Hayden Planetarium. In the summer of 2002, by which time Tyson had become the Hayden's director, the museum's shrinking budget and changing vision led to the placement of the magazine in private hands. That's when I became a senior editor at *Natural History* and, more specifically, Tyson's editor—a relationship still in force, though both of us have now, separately, moved on from the magazine.

You wouldn't think an erstwhile art historian and curator would be the ideal editor for Tyson. But here's the thing: he cares about communication, he cares about fostering science literacy, and if, together, we can produce something that I comprehend and that sounds good to him, then we've both succeeded.

It's been more than half a century since the Soviet Union put a small, beeping metal sphere into Earth orbit, and not much less than half a century since the United States sent its first astronauts for a stroll on the Moon. A wealthy individual can now book a personal trip to space for $20 million or $30 million. Private US aerospace companies are testing vehicles suitable for ferrying crew and cargo to and from the International Space Station. Satellites are becoming so numerous that geosynchronous orbit is almost running out of room. Tallies of wayward orbital debris larger than half an inch now number in the hundreds of thousands. There is talk of mining asteroids and concern about the militarization of space.

During the opening decade of the present century in America, blue-ribbon commissions and reports initially fostered dreams not only of a swift US manned return to the Moon but of more distant human space travel as well. NASA's budgets have not matched its mandates, however, and so its recent achievements beyond Earth's atmosphere have involved human activities only within low Earth orbit, and only robotic activities at greater distances. In early 2011 NASA warned Congress that neither prevalent launch-system designs nor customary funding levels are capable of getting the United States back to space by 2016.

Meanwhile, other countries have hardly been asleep at the wheel. China sent up its first astronaut in 2003; India plans to do the same in 2015. The European Union sent its first probe to the Moon in 2004; Japan sent its first in 2007; India sent its first in 2008. On October 1, 2010, the sixty-first National Day of the People's Republic, China carried out a flawless launch of its second unmanned Moon probe, whose job is to survey possible landing sites for China's third Moon probe. Russia, too, is planning a return visit. Brazil, Israel, Iran, South Korea, and Ukraine, as well as Canada, France, Germany, Italy, and the UK, all have firmly established, highly active space agencies. Some four dozen countries operate satellites. South Africa has just formed a national space agency; someday there will be a pan-Arab space agency. Multinational collaboration is becoming de rigueur. Beyond as well as within America, most of the world's scientists recognize that space is a global commons—a domain appropriate only for collectivity—and they expect collective progress to continue despite crises, limitations, and setbacks.

Neil deGrasse Tyson has thought, written, and spoken about all these things and many more. In this volume we have collected fifteen years' worth of his commentaries on space exploration, organizing them within what seemed to us an organic framework: Part I—"Why," Part II—"How," and Part III—"Why Not." Why does the human animal wonder about space, and why must we explore it? How have we managed to reach space thus far, and how might we reach it in the future? What obstacles prevent the fulfillment of the space enthusiasts' daring dreams? A dissection of the politics of space opens the anthology; a deliberation on the meaning of space completes it. At the very end are indispensable appendices: the text of the National Aeronautics and Space Act of 1958; extracts of related legislation; charts showing the space budgets of multiple US government agencies and multiple countries, as well as the trajectory of NASA spending over the course of half a century in relation to total federal spending and the overall US economy.

Eventually, if not as astronauts then as atoms, we'll all be caught up in the blizzard of icy dust, the electromagnetic radiation, the soundlessness and peril that constitute space. Right now, though, Tyson is onstage, ready to usher us through catastrophes one minute and crack us up the next. Listen up, because living off-planet might lie ahead.

AVIS LANG

SPACE
CHRONICLES

PROLOGUE

Space Politics

> You develop an instant global consciousness, a people orientation, an intense dissatisfaction with the state of the world, and a compulsion to do something about it. From out there on the moon, international politics look so petty. You want to grab a politician by the scruff of the neck and drag him a quarter of a million miles out and say, "Look at that!"
>
> —Edgar Mitchell, Apollo 14 astronaut, 1974

Some people think emotionally more often than they think politically. Some think politically more often than they think rationally. Others never think rationally about anything at all.

No judgment implied. Just an observation.

Some of the most creative leaps ever taken by the human mind are decidedly irrational, even primal. Emotive forces are what drive the greatest artistic and inventive expressions of our species. How else could the sentence "He's either a madman or a genius" be understood?

It's okay to be entirely rational, provided everybody else is too. But apparently this state of existence has been achieved only in fiction, as in the case of the Houyhnhnms, the community of intelligent horses that Lemuel Gulliver stumbles upon during his early eighteenth-century travels (the name "Houyhnhnm" translates from the local language as "perfection of nature"). We also find a rational society among the Vulcan race in the perennially popular science-fiction series *Star Trek*. In both worlds, societal decisions get made with efficiency and dispatch, devoid of pomp, passion, and pretense.

To govern a society shared by people of emotion, people of reason, and everybody in between—as well as people who think their actions are shaped by logic but in fact are shaped by feelings or nonempirical philosophies—you need politics. At its best, politics navigates all these mind-states for the sake of the greater good, alert to the rocky shoals of community, identity, and the economy. At its worst, politics thrives on the incomplete disclosure and misrepresentation of data required by an electorate to make informed decisions, whether arrived at logically or emotionally.

On this landscape we find intractably diverse political views, with no obvious hope of consensus or even convergence. Some of the hottest of the hot-button issues include abortion, capital punishment, defense spending, financial regulation, gun control, and tax laws. Where you stand on these issues correlates strongly with your political party's portfolio of beliefs. In some cases it's more than correlation; it's the foundation of a political identity.

All this may leave you wondering how anything productive can ever happen under a politically fractious government. Credit comedian Gallagher, in his 1985 film *The Bookkeeper*, with the observation that if con is the opposite of pro, then Congress must be the opposite of progress.

Until recently, space exploration stood above party politics. NASA was more than bipartisan; it was nonpartisan. Specifically, a person's support for NASA was uncorrelated with whether or not that person was liberal or conservative, Democrat or Republican, urban or rural, impoverished or wealthy.

NASA's placement in American culture further bears this out. The ten NASA centers are geographically distributed across eight states. Following the 2008 federal election, they were represented in the House by six Democrats and four Republicans; in the 2010 election that distribution was reversed. Senators from those states are similarly balanced, with eight Republicans and eight Democrats. This "left-right" representation has been a persistent feature of NASA's support over the years. The National Aeronautics and Space Act of 1958 became law under Republican president Dwight D. Eisenhower. Democratic president John F. Kennedy launched the Apollo program in 1961. Republican president Richard M. Nixon's signature is on the plaque left on the Moon in 1969 by the Apollo 11 astronauts.

And maybe it's just coincidence, but twenty-four astronauts hail from the swing state of Ohio—more than from any other state—including John Glenn (America's first to orbit Earth) and Neil Armstrong (the world's first to walk on the Moon).

If partisan politics ever leaked into NASA's activities, it tended to appear on the fringes of operations. For example, President Nixon could, in principle, have dispatched the newly commissioned USS John F. Kennedy aircraft carrier to pluck the Apollo 11 command module from the Pacific Ocean. That would have been a nice touch. Instead he sent the USS Hornet, a more expedient option at the time. The Kennedy never saw the Pacific, and was in dry dock in Portsmouth, Virginia, for the July 1969 splashdown. Consider another example: With top cover from the industry-friendly Republican president Ronald Reagan, Congress passed the Commercial Space Launch Act of 1984, which not only allowed but also encouraged civilian access to NASA-funded innovations related to launch vehicles and space hardware, thereby opening the space frontier to the private sector. A Democrat might or might not have thought up that legislation, but a Republican Senate and a Democratic House of Representatives both passed it, and the concept is as American as a moonwalk.

One could further argue that NASA's achievements transcend nations. Stunning images of the cosmos from the Hubble Space Telescope have brought the distant universe into focus for everyone with an Internet connection. Apollo astronauts have appeared on postage stamps from other countries, including Dubai and Qatar. And in the 2006 documentary *In the Shadow of the Moon,* Apollo 12 astronaut Alan Bean, the fourth person to walk on the Moon, comments that during his international travels people would jubilantly declare, "We did it!" They didn't say, "You did it!" or "America did it!" The moonwalkers, though 83 percent military and 100 percent American male, were emissaries of our species, not of a nation or political ideology.

Although NASA has historically been free from partisanship, it's been anything but free from politics itself, driven especially by international forces much greater than any purely domestic initiatives can muster. With the 1957 Soviet launch of Sputnik 1, the world's first artificial satellite, America was spooked into the space race. A year later, NASA itself was birthed in a climate of Cold War fears. Mere weeks after the Soviets put the first person into orbit, the United States was spooked into creating the Apollo program to the Moon. Over that time, the Soviet Union beat us in practically every important measure of space achievement: first spacewalk, longest spacewalk, first woman in space, first docking in space, first space station, longest time logged in space. By declaring the race to be about reaching the Moon and nothing else, America gave itself permission to ignore the contests lost along the way.

Having beaten the Russians to the Moon, we declare victory and—with no chance of their putting a person on the lunar surface—we stop going there altogether. What happens next? The Russians "threaten" to build massive space platforms equipped to observe all that happens on Earth's surface. This decades-long effort, which begins in 1971 with a series of Salyut (Russian for "salute") space modules, culminates with space station Mir (Russian for "peace"), the world's first permanently inhabited space platform, whose assembly began in 1986. Once again, being reactive rather than proactive to geopolitical forces, America concludes that we need one of those too. In his 1984 State of the Union address, President Reagan announces rather urgent plans to design and build Space Station Freedom, with nations friendly to our politics joining the effort. Though approved by Congress, the project's full scope and expense does not survive 1989, the year that peace breaks out in Europe as the Cold War draws to a close. President Clinton collects the underfunded pieces and, by 1993, puts into play a reconceived platform—the International Space Station (some assembly required)—that calls for the participation of former arch-enemy Russia. This strategic move offers wayward Russian nuclear scientists and engineers something interesting to do other than make weapons of mass destruction for our emergent adversaries around the globe. That same year would see the cancellation of the Superconducting Super Collider, an expensive physics experiment that had been approved in the 1980s during a Cold War Congress. Unaffordable cost overruns are the reason usually cited for the cancellation, but one cannot ignore the politically abrasive fact that the space station and the collider would both be managed in Texas, amounting to more pork than any state deserves in a single budget cycle. History, however, offers an even deeper reason. In peacetime, the collider did not enjoy the same strategic value to America's national security as did the space station. Once again, politics and war trumped the urge to discover.

Other than military alliances, the International Space Station remains one of the most successful collaborations of countries. Besides Russia, participating members include Canada, Japan, Brazil, and eleven member nations of the European Space Agency: Belgium, Denmark, France, Germany, Italy, the Netherlands, Norway, Spain, Sweden, Switzerland, and the United Kingdom. Citing human rights violations, we exclude China from this collaboration. But that's not enough to stymie an ambitious country. Undeterred, China births an independent manned space program, launching Yang Liwei as its first taikonaut in 2003. Like the first American astronauts, Yang was a fighter pilot. The choice of Yang, together with other posturings within China's space program, such as the kinetic kill of a defunct but still-orbiting weather satellite by a medium-range ballistic missile, causes some American analysts to see China as an adversary, with the capacity to threaten US access to space as well as US assets that reside there.

Wouldn't it be a curious twist of events if China's vigorous response to our denial of their participation in the International Space Station turns out to be the very force that sparks another series of competitive space achievements in America, culminating this time around in a manned mission to Mars?

Averaged over its history, NASA spends about $100 billion in today's dollars every five or six years. Hardly anywhere in that stream of money have NASA's most expensive initiatives (including the Mercury, Gemini, and Apollo programs, propulsion research, the space shuttle, and the space station) been driven by science or discovery or the betterment of life on Earth. When science does advance, when discovery does unfold, when life on Earth does improve, they happen as an auxiliary benefit and not as a primary goal of NASA's geopolitical mission.

Failure to embrace these simple realities has led to no end of delusional analysis of what NASA is about, where NASA has been, and where NASA will likely ever go.

On July 20, 1989, twenty years to the day after the Apollo 11 Moon landing, President George Bush Sr. delivered a speech at the National Air and Space Museum, using the auspicious anniversary to announce the Space Exploration Initiative. It reaffirmed the need for Space Station Freedom, but also called for a permanent presence on the Moon and a manned voyage to Mars. Invoking Columbus, the president likened his plan to epic episodes of discovery in the history of nations. He said all the right things, at the right time and the right place. So how could the stirring rhetoric not have worked? It worked for President Kennedy on September 12, 1962, at Rice University Stadium in Houston. That's when and where he described what would become the Apollo program, declaring, with politically uncommon fiscal candor: "To be sure, all this costs us all a good deal of money. This year's space budget is three times what it was in January 1961, and it is greater than the space budget of the previous eight years combined."

Maybe all Bush needed was some of that famous charisma that Kennedy exuded. Or maybe he needed something else.

Shortly after Bush's speech, a group led by the director of NASA's Johnson Space Center presented a cost analysis for the entire plan that reported a coffer-constricting, Congress-choking price tag of $500 billion over twenty to thirty years. The Space Exploration Initiative was dead on arrival. Was it any more costly than what Kennedy asked for, and got? No. It was less. Not only that, since $100 billion over five or six years represents NASA's baseline funds, thirty years of that spending level gets you to the $500 billion mark without ever having to top up the budget.

The opposite outcomes of these two speeches had nothing do with political will, public sentiment, persuasiveness of arguments, or even cost. President Kennedy was at war with the Soviet Union, whereas President Bush wasn't at war with anybody. When you're at war, money flows like a tapped keg, rendering irrelevant the existence or absence of other variables, charisma included.

Meanwhile, space zealots who do not properly factor the role of war into the spending landscape are delusionally certain that all we need today are risk-taking visionaries like JFK. Couple that with the right dose of political will, they contend, and we surely would have been on Mars long ago, with hundreds if not thousands of people living and working in space colonies. Princeton space visionary Gerard K. O'Neill, among others, imagined all this in place by the year 2000.

The opposite of space zealots—space curmudgeons—are those who are certain that NASA is a waste of taxpayer money and that funds allocated via NASA centers are the equivalent of pork-barrel spending. Genuine pork, of course, is money procured by individual members of Congress for the exclusive benefit of their own districts, with no tangible gain to any other. NASA, by and large, is the opposite of this. The nation and the world thrive on NASA's regional innovations, which have transformed how we live.

Here's an experiment worth conducting. Sneak into the home of a NASA skeptic in the dead of night and remove all technologies from the home and environs that were directly or indirectly influenced by space innovations: microelectronics, GPS, scratch-resistant lenses, cordless power tools, memory-foam mattresses and head cushions, ear thermometers, household water filters, shoe insoles, long-distance telecommunication devices, adjustable smoke detectors, and safety grooving of pavement, to name a few. While you're at it, make sure to reverse the person's LASIK surgery. Upon waking, the skeptic embarks on a newly barren existence in a state of untenable technological poverty, with bad eyesight to boot, while getting rained on without an umbrella because of not knowing the satellite-informed weather forecast for that day.

When NASA's manned missions are not advancing a space frontier, NASA's science activities tend to dominate the nation's space headlines, which currently emanate from four divisions: Earth Science, Heliophysics, Planetary Science, and Astrophysics. The largest portion of NASA's budget ever spent on these activities briefly hit 40 percent, in 2005. During the Apollo era, the annual percentage hovered in the mid-teens. Averaged over NASA's half century of existence, the annual percentage of spending on science sits in the low twenties. Put simply, science is not a funding priority either for NASA or for any of the members of Congress who vote to support NASA's budget.

Yet the word "science" is never far from the acronym "NASA" in anybody's discussion of why NASA matters. As a result, even though geopolitical forces drive spending on space exploration, exploring space in the name of science plays better in public discourse. This mismatch of truth and perceived truth leads to two outcomes. In speeches and testimonies, lawmakers find themselves overstating the actual scientific return on manned

NASA missions and programs. Senator John Glenn, for instance, has been quick to celebrate the zero-G science potential of the International Space Station. But with its budget of $3 billion per year, is that how a community of researchers would choose to spend the cash? Meanwhile, in the academic community, pedigreed scientists heavily criticize NASA whenever money is spent on exploration with marginal or no scientific return. Among others of that sentiment, the particle physicist and Nobel laureate Steven Weinberg is notably blunt in his views, expressed, for example, in 2007 to a Space.com reporter during a scientific conference at Baltimore's Space Telescope Science Institute:

> The International Space Station is an orbital turkey. . . . No important science has come out of it. I could almost say no science has come out of it. And I would go beyond that and say that the whole manned spaceflight program, which is so enormously expensive, has produced nothing of scientific value.
>
> . . . NASA's budget is increasing, with the increase being driven by what I see on the part of the president and the administrators of NASA as an infantile fixation on putting people into space, which has little or no scientific value.

Only those who believe deep down that NASA is (or should be) the exclusive private funding agency of scientists could make such a statement. Here's another: an excerpt from the resignation letter of Donald U. Wise, NASA's chief lunar scientist. Though less acerbic than Weinberg's statement, it shares a kindred spirit:

> I watched a number of basic management decisions being made, shifting priorities, funds and manpower away from maximization of exploration capabilities . . . toward the development of large new manned space systems.
>
> Until such time as [NASA] determines that science is a major function of manned space flight and is to be supported with adequate manpower and funds, any other scientist in my vacated position would also be likely to expend his time futilely.

With these comments submitted as evidence, one might suppose that NASA's interest in science has ebbed since the old days. But Wise's letter is, in fact, from the old days: August 24, 1969, thirty-five days after we first stepped foot on the Moon.*

What an ivory-tower luxury it is to lament that NASA is spending too little on science. Unimagined in these complaints is the fact that without geopolitical drivers, there would likely be no NASA science at all.

America's space program, especially the golden era of Apollo and its influence on the dreams of a nation, makes fertile rhetoric for almost any occasion. Yet the deepest message therein is often neglected, misapplied, or forgotten altogether. In a speech delivered at the National Academy of Sciences on April 27, 2009, President Barack Obama waxed poetic about NASA's role in driving American innovation:

> President Eisenhower signed legislation to create NASA and to invest in science and math education, from grade school to graduate school. And just a few years later, a month after his address to the 1961 Annual Meeting of the National Academy of Sciences, President Kennedy boldly declared before a joint session of Congress that the United States would send a man to the moon and return him safely to the earth.
>
> The scientific community rallied behind this goal and set about achieving it. And it would not only lead to those first steps on the moon; it would lead to giant leaps in our understanding here at home. That Apollo program produced technologies that have improved kidney dialysis and water purification systems; sensors to test for hazardous gases; energy-saving building materials; fire-resistant fabrics used by firefighters and soldiers. More broadly, the enormous investment in that era—in science and technology, in education and research funding—produced a great outpouring of curiosity and creativity, the benefits of which have been incalculable.

What's stunning about Obama's message is that the point of his speech was to alert the academy to the proposed American Recovery and Reinvestment Act—legislation that would place the budgets of the National

* Letter from Donald Wise, chief scientist and deputy director, Apollo Lunar Exploration Office, NASA, to Homer Newell, associate administrator, NASA, August 24, 1969. Reprinted in John M. Logsdon et al., eds., *Exploring the Unknown: Selected Documents in the History of the U.S. Civil Space Program*, vol. 5: *Exploring the Cosmos*, NASA SP-2001-4407 (Washington, DC: Government Printing Office, 2001), 185–86.

Science Foundation, the Department of Energy's Office of Science, and the National Institute of Standards and Technology on a path to double over the coming years. Surely NASA's budget would be doubled too? Nope. All NASA got was a single-year allocation of a billion dollars. Given that space exploration formed the rhetorical soul of the president's speech, this move defies rational, political, and even emotional analysis.

For his second State of the Union Address, delivered January 27, 2011, President Obama once again cited the space race as a catalyst for scientific and technological innovation. That original "Sputnik moment"—crystallized in Kennedy's 1961 speech to a joint session of Congress—is what got us to the Moon and set the highest of bars for America's vision and leadership in the twentieth century. As the president rightly recounted, "We unleashed a wave of innovation that created new industries and millions of new jobs." Citing the hefty investments that other countries are making in their technological future, and the tandem failing of America's educational system to compete on the world stage, Obama declared the disturbing imbalance to be this generation's Sputnik moment. He then challenged us by 2015 to (1) have a million electric vehicles on the road and (2) deploy the next generation of high-speed wireless to 98 percent of all Americans—and by 2035 to (1) derive 80 percent of America's electricity from clean energy and (2) give 80 percent of Americans access to high-speed rail.

Laudable goals, all of them. But to think of that list as the future fruits of a contemporary Sputnik moment dispirits the space enthusiast. It reveals a change of vision over the decades, from dreams of tomorrow to dreams of technologies that should already have been with us.

Following the February 1, 2003, loss of the Columbia space shuttle orbiter and its crew of seven, the public and press, as well as key lawmakers, called for a new NASA vision—one with its sights set beyond low Earth orbit. What better time to reassess a program than after a disaster? Makes you wonder, however, why the Challenger disaster in 1986, which also resulted in the loss of a seven-person crew, did not trigger a similar call for a renewed NASA mission statement. Why? In 1986, nothing much was happening in the Chinese space community. By contrast, on October 15, 2003, China launched its first taikonaut into Earth orbit, becoming just the third nation to join the spacefarers' club.

A mere three months later, on January 15, 2004, the Bush White House announced a brand-new Vision for Space Exploration. The time had finally arrived for the United States to leave low Earth orbit again.

The vision was a basically sound plan that also called for completion of the International Space Station and retirement of NASA's space shuttle workhorse by decade's end, with the recovered funds used to create a new launch architecture that would take us back to the Moon and onward to more distant places. But beginning in February 2004 (with my appointment by President Bush to the nine-member Commission on Implementation of United States Space Exploration Policy, whose mandate was to chart an affordable and sustainable course of action), I began to notice a pall of partisanship descending on NASA and on the nation's space policy. Strong party allegiances were clouding, distorting, and even blinding people's space sensibilities across the entire political spectrum.

Some Bush-bashing Democrats, predisposed to think politically rather than rationally, were quick to criticize the plan on the grounds that the nation could not afford it, even though our commission was explicitly charged with keeping costs in check. Other Democrats argued that the space vision offered no details regarding its implementation. Yet supportive documents were freely available from the White House and from NASA. Consider also that President Bush delivered his speech on the plan at NASA's DC headquarters. No sitting president had ever done such a thing. To cover the West Coast, Bush tasked Vice President Cheney to speak at NASA's Jet Propulsion Laboratories in Pasadena, California, on the same day. (By way of comparison, President Kennedy's May 25, 1961, address to a joint session of Congress contains only a couple of paragraphs urging that a Moon mission be funded.) Other disgruntled Democrats, still fulminating about the controversial election in 2000 and feeling deep dissatisfaction with Bush's first term in office, commonly quipped that we should instead send Bush to Mars.

All told, the criticisms were not only underinformed but also betrayed a partisan bias I hadn't previously encountered during my years of exposure to space politics—although I am happy to report that after all the knee-jerk reactions ran their course, the 2004 Vision for Space Exploration secured strong bipartisan support.

With Barack Obama in office beginning in 2009, the level of vitriol from extreme Republicans exceeded even that of the extreme Democrats who found nothing praiseworthy in anything President Bush ever said, thought, or did. On April 15, 2010, Obama delivered a space policy speech at the Kennedy Space Center in Florida that I happened to attend. Factoring out Obama's Kennedyesque charisma and undeniable oratorical skills, I can objectively say that he delivered a powerful, hopeful message for the future of America's space exploration—a vision that would lead us to multiple places beyond low Earth orbit, asteroids included. He also reaffirmed the need to retire the space shuttle and spoke longingly of Mars. President Obama even went one step further, suggesting that since we've already been to the Moon, why return at all? Been there, done that. With an advanced launch vehicle—one that leapfrogs previous rocket technologies but would take many years to develop—we could bypass the Moon altogether and head straight for Mars by the mid-2030s, right about when Obama expects 80 percent of Americans to abandon cars and planes, and instead travel to and fro via high-speed rail.

I was there. I felt the energy of the room. More important, I resonated with Obama's enthusiasm for NASA and its role in shaping the American zeitgeist. As for coverage of the speech, a typical headline in the Obama-supportive press was "Obama Sets Sights on Mars." The Obama-resenting press, however, declared: "Obama Kills Space Program." You can't get more partisan than that.

Scores of protesters lined the Kennedy Space Center's surrounding causeways that day, wielding placards that pleaded with the president not to destroy NASA. In the weeks to follow, many people—including marquee astronauts—felt compelled to choose sides. Two moonwalkers sharply critical of Obama's plan to cancel the return to the Moon testified before Congress: Neil Armstrong of Apollo 11 and Eugene Cernan of Apollo 17, poignantly presented as the first and the last to step foot on the Moon. On the other hand, Neil Armstrong's command-module partner Buzz Aldrin was strongly supportive of Obama's plan and had accompanied the president to Florida aboard Air Force One.

Either Obama had given two different speeches at the Kennedy Space Center that morning and I heard only one of them, or else everyone in the room (myself included, perhaps) was suffering from a bad case of selective hearing.

Indeed, the president did deliver more than one speech that day—or rather, his single coherent plan had different consequences for different people. As an academic with a long-term view, I focused on Obama's thirty-year vision for NASA, and I celebrated it. But to somebody who wants uninterrupted access to space, in their own country's launch vehicle, controlled by their own country's astronauts, any halt to our space access is simply unacceptable. It's worth remembering that during the halt in shuttle launches that followed the Columbia tragedy, the Russians were happy to "shuttle" our astronauts back and forth to the International Space Station aboard their reliable Soyuz capsule. So the stipulation that American access to orbit shall always and forever be in a craft of our own manufacture may be an example of pride overriding practicality. And by the way, there was barely a peep back in 2004 when President Bush first proposed to phase out the shuttle. President Obama was simply following through on Bush's plan.

Taken at face value, the opposite reactions to Obama's words need not reflect a partisan divide. They could simply be honest differences of opinion. But they weren't. Views and attitudes split strongly along party lines, requiring olive-branch compromises in Congress before any new budget for NASA could be agreed upon and passed. A letter I was invited to submit to lawmakers—reaffirming NASA's value to America's identity and future while also urging a swift solution to the impasse—became a twig on one of those olive branches. A bipartisan posse of solution-seeking congressmen attempted to alter the president's proposal and the associated budget for NASA in a way that would appease the fundamentally Republican-led resistance. They sought to accelerate the design and construction of the heavy-lift launch architecture that would enable the first manned mission beyond low Earth orbit since the Apollo era's Saturn V rocket. This deceptively simple adjustment to the plan would help close the gap between the twilight of America's shuttle launches and the dawn of a new era of launch capability—and, as a consequence, preserve aerospace jobs that the Obama plan would have destabilized.

Jobs? Is that what it's about? Now it all made sense. I'd thought the real issue was the cultural imperative of continuous access to space and the short-term fate of the manned program. Surely that's what all the protest placards meant, as well as the associated anti-Obama rhetoric. But if jobs are what really matters to everybody, why don't they just say so? If I were a shuttle worker at any level—especially if I were a contractor to NASA in support of launch operations—then the gap between the phaseout of the shuttle and the next rocket to launch beyond Earth is all I would have heard in the president's speech. And if new, nonderivative, uncertain launch technologies would be required to achieve the vision, then the downtime for manned space flight in America would also be uncertain, which means the only thing certain in the face of these uncertainties is that I'd be out of a job.

Since the shuttle is a major part of NASA operations, and NASA's industrial partners are spread far and wide across the American countryside, an unemployment ripple gets felt in many more places than the causeways of Florida's Space Coast. President Obama's speech did include mention of funded retraining programs for workers whose jobs would be eliminated. He also noted that his plan would erase fewer jobs than his predecessor's Vision for Space Exploration would have done—had it been implemented—although he put a positive spin on that fact by asserting, "Despite some reports to the contrary, my plan will add more than 2,500 jobs along the Space Coast in the next two years compared to the plan under the previous administration."

That line received immediate applause. I wonder what the reaction in the room would have been if Obama's statement were mathematically equivalent but more blunt: "Bush's plan would have destroyed 10,000 jobs; my plan would destroy only 7,500."

Applause notwithstanding, Obama's message fell flat in the hearts and minds of entire corps of skilled technologists who had forged their multidecade careers on doing whatever it took to get the shuttle into orbit. So anybody who didn't like President Obama before the speech at the Kennedy Space Center now had extra reasons to brand him as the villain: In 1962 there were two spacefaring nations. Fifty years later, in 2012, there would still be two spacefaring nations. But America wouldn't be one of them.

It's now retrospectively obvious why nary a mention of jobs appeared in the anti-Obama protest mantras: nobody but nobody, especially a Republican, wants to be thought of as someone who sees NASA as a government jobs program, although that comment has been made before—not by a politician, but by a comedian. The always candid and occasionally caustic Wanda Sykes allots two full, disdainful pages of her 2004 book *Yeah, I Said It* to NASA's exploits. On the subject of jobs: "NASA is a billion dollar welfare program for really smart dorks. Where else are they going to work? They're too smart to do anything else."

Among the reasons one might take issue with Obama's space vision, there's a far deeper one than the ebb and flow of jobs. In an electoral democracy, a president who articulates any goal for which the completion lies far beyond his tenure cannot guarantee ever reaching that goal. In fact, he can barely guarantee reaching any goal whatsoever during his time in office. As for goals that activate partisan sensitivities, a two-term president faces the additional risk of multiple biennial shifts in the ruling parties of Congress.

Kennedy knew full well what he was doing in 1961 when he set forth the goal of sending a human to the Moon "before this decade is out." Had he lived and been elected to a second term, he would have been president through January 19, 1969. And had the Apollo 1 launchpad fire that killed three astronauts not delayed the program, we would certainly have reached the Moon during his presidency.

Now imagine, instead, if Kennedy had called for achieving the goal "before this century is out." With that as a vision statement, it's not clear whether we would have ever left Earth. When a president promises something beyond his presidency, he's fundamentally unaccountable. It's not his budget that must finish the job. It becomes another president's inherited problem—a ball too easily dropped, a plan too easily abandoned, a dream too readily deferred. So while the rhetoric of Obama's space speech was brilliant and visionary, the politics of his speech were, empirically, a disaster. The only thing guaranteed to happen on his watch is the interruption of America's access to space.

Every several years for the past several decades, NASA gets handed a "new direction." Many different factions within the electorate believe they know what's right for the agency as they fight one another over its future. The only good part about these battles, enabling hope to spring eternal, is that hardly anybody is arguing about whether NASA should exist at all—a reminder that we are all stakeholders in our space agency's uncertain future.

Collectively, the selections in this volume investigate what NASA means to America and what space exploration means to our species. Although the path to space is scientifically straightforward, it is nonetheless technologically challenging and, on too many occasions, politically intractable. Solutions do exist. But to arrive at them, we must abandon delusional thinking and employ tools of cultural navigation that link space exploration with science literacy, national security, and economic prosperity. Thus equipped, we can invigorate the nation's mandate to compete internationally while at the same time fueling the timeless urge to discover what lies beyond the places we already know.

PART I

WHY

••• CHAPTER ONE

THE ALLURE OF SPACE*

For millennia, people have looked up at the night sky and wondered about our place in the universe. But not until the seventeenth century was any serious thought given to the prospect of exploring it. In Proposition 14 of a charming book published in 1640, *The Discovery of a World in the Moone,* the English clergyman and science buff John Wilkins speculates on what it might take to travel in space:

> [Y]et I do seriously, and upon good grounds, affirm it possible, to make a flying chariot, in which a man may sit and give such a motion unto it as shall convey him through the air; and this, perhaps, might be made large enough to carry divers men at the same time. . . . We see a great ship swim as well as a small cork; and an eagle flies in the air as well as a little gnat. . . . So that notwithstanding all [the] seeming impossibilities, tis likely enough there may be a means invented of journeying to the Moon; and how happy they shall be that are first successful in this attempt.

Three hundred and twenty-nine years later, humans would indeed land on the Moon, aboard a chariot called Apollo 11, as part of an unprecedented investment in science and technology conducted by a relatively young country called the United States of America. That enterprise drove a half century of unprecedented wealth and prosperity that today we take for granted. Now, as our interest in science wanes, America is poised to fall

* Adapted from "Why America Needs to Explore Space," *Parade,* August 5, 2007.

behind the rest of the industrialized world in every measure of technological proficiency.

In recent decades, the majority of students in America's science and engineering graduate schools have been foreign-born. Up through the 1990s, most would come to the United States, earn their degrees, and gladly stay here, employed in our high-tech workforce. Now, with emerging economic opportunities back in India, China, and Eastern Europe—the regions most highly represented in advanced academic science and engineering programs—many graduates choose to return home.

It's not a brain drain—because America never laid claim to these students in the first place—but a kind of brain regression. The slow descent from America's penthouse view, enabled by our twentieth-century investments in science and technology, has been masked all these years by self-imported talent. In the next phase of this regression we will begin to lose the talent that trains the talent. That's a disaster waiting to happen; science and technology are the greatest engines of economic growth the world has seen. Without regenerating homegrown interest in these fields, the comfortable lifestyle to which Americans have become accustomed will draw to a rapid close.

Before visiting China in 2002, I had pictured a Beijing of wide boulevards, dense with bicycles as the primary means of transportation. What I saw was very different. Of course the boulevards were still there, but they were filled with top-end luxury cars; construction cranes were knitting a new skyline of high-rise buildings as far as the eye could see. China has completed the controversial Three Gorges Dam on the Yangtze River, the largest engineering project in the world—generating more than twenty times the energy of the Hoover Dam. It has also built the world's largest airport and, as of 2010, had leapfrogged Japan to become the world's second-largest economy. It now leads the world in exports and CO_2 emissions.

In October 2003, having launched its first taikonaut into orbit, China became the world's third spacefaring nation (after the United States and Russia). Next step: the Moon. These ambitions require not only money but also people smart enough to figure out how to turn them into reality, and visionary leaders to enable them.

In China, with a population approaching 1.5 billion, if you are smart enough to be one in a million, then there are 1,500 other people just like you.

Meanwhile, Europe and India are redoubling their efforts to conduct robotic science on spaceborne platforms, and there's a growing interest in space exploration from more than a dozen other countries around the world, including Israel, Iran, Brazil, and Nigeria. China is building a new space launch site whose location, just nineteen degrees north of the equator, makes it geographically better for most space launches than Cape Canaveral is for the United States. This growing community of space-minded nations is hungry for its slice of the aerospace universe. In America, contrary to our self-image, we are no longer leaders, but simply players. We've moved backward just by standing still.

Space Tweet #1
100,000: Altitude, in meters, above Earth's surface where International Federation of Aeronautics defines beginning of space
Jan 23, 2011 9:47 AM

Tweet photo by Dan Deitch © WGBH Educational Foundation

But there's still hope for us. You can learn something deep about a nation when you look at what it accomplishes as a culture. Do you know the most popular museum in the world over the past decade? It's not the Metropolitan Museum of Art in New York. It's not the Uffizi in Florence. It's not the Louvre in Paris. At a running average of some nine million visitors per year, it's the National Air and Space Museum in Washington, DC, which contains everything from the Wright Brothers' original 1903 aeroplane to the Apollo 11 Moon capsule, and much, much more. International visitors are anxious to see the air and space artifacts housed in this museum, because they're an American legacy to the world. More important, NASM represents the urge to dream and the will to enable it. These traits are fundamental to being human, and have fortuitously coincided with what it has meant to be American.

When you visit countries that don't nurture these kinds of ambitions, you can feel the absence of hope. Owing to all manner of politics, economics, and geography, people are reduced to worrying only about that day's shelter or the next day's meal. It's a shame, even a tragedy, how many people do not get to think about the future. Technology coupled with wise leadership not only solves these problems but enables dreams of tomorrow.

For generations, Americans have expected something new and better in their lives with every passing day—something that will make life a little more fun to live and a little more enlightening to behold. Exploration accomplishes this naturally. All we need to do is wake up to this fact.

The greatest explorer of recent decades is not even human. It's the Hubble Space Telescope, which has offered everybody on Earth a mind-expanding window to the cosmos. But that hasn't always been the case. When it was launched in 1990, a blunder in the manufacture of the optics generated hopelessly blurred images, much to everyone's dismay. Three years would pass before corrective optics were installed, enabling the sharp images that we now take for granted.

What to do during the three years of fuzzy images? It's a big, expensive telescope. Not wise to let it orbit idly. So we kept taking data, hoping some useful science would nonetheless come of it. Eager astrophysicists at Baltimore's Space Telescope Science Institute, the research headquarters for the Hubble, didn't just sit around; they wrote suites of advanced image-processing software to help identify and isolate stars in the otherwise crowded, unfocused fields the telescope presented to them. These novel techniques allowed some science to get done while the repair mission was being planned.

Meanwhile, in collaboration with Hubble scientists, medical researchers at the Lombardi Comprehensive Cancer Center at Georgetown University Medical Center in Washington, DC, recognized that the challenge faced by astrophysicists was similar to that faced by doctors in their visual search for tumors in mammograms. With the help of funding from the National Science Foundation, the medical community adopted these new techniques to assist in the early detection of breast cancer. That means countless women are alive today because of ideas stimulated by a design flaw in the Hubble Space Telescope.

You cannot script these kinds of outcomes, yet they occur daily. The cross-pollination of disciplines almost always creates landscapes of innovation and discovery. And nothing accomplishes this like space exploration, which draws from the ranks of astrophysicists, biologists, chemists, engineers, and planetary geologists, whose collective efforts have the capacity to improve and enhance all that we have come to value as a modern society.

How many times have we heard the mantra "Why are we spending billions of dollars up there in space when we have pressing problems down here on Earth?" Apparently, the rest of world has no trouble coming up with good answers to this question—even if we can't. Let's re-ask the question in an illuminating way: "As a fraction of your tax dollar today, what is the total cost of all spaceborne telescopes, planetary probes, the rovers on Mars, the International Space Station, the space shuttle, telescopes yet to orbit, and missions yet to fly?" Answer: one-half of one percent of each tax dollar. Half a penny. I'd prefer it were more: perhaps two cents on the dollar. Even during the storied Apollo era, peak NASA spending amounted to little more than four cents on the tax dollar. At that level, the Vision for Space Exploration would be sprinting ahead, funded at a level that could reclaim our preeminence on a frontier we pioneered. Instead the vision is just ambling along, with barely enough support to stay in the game and insufficient support ever to lead it.

So with more than ninety-nine out of a hundred cents going to fund all the rest of our nation's priorities, the space program does not prevent (nor has it ever prevented) other things from happening. Instead, America's former investments in aerospace have shaped our discovery-infused culture in ways that are obvious to the rest of the world, whether or not we ourselves recognize them. But we are a sufficiently wealthy nation to embrace this investment in our own tomorrow—to drive our economy, our ambitions, and, above all, our dreams.

••• CHAPTER TWO

EXOPLANET EARTH*

Whether you prefer to crawl, sprint, swim, or walk from one place to another, you can enjoy close-up views of Earth's inexhaustible supply of things to notice. You might see a vein of pink limestone on the wall of a canyon, a ladybug eating an aphid on the stem of a rose, a clamshell poking out of the sand. All you have to do is look.

Board a jetliner crossing a continent, though, and those surface details soon disappear. No aphid appetizers. No curious clams. Reach cruising altitude, around seven miles up, and identifying major roadways becomes a challenge.

Detail continues to vanish as you ascend to space. From the window of the International Space Station, which orbits at about 225 miles up, you might find London, Los Angeles, New York, or Paris in the daytime, not because you can see them but because you learned where they are in geography class. At night, brightly lit megacities present as patches of glow. By day, contrary to common wisdom, with the unaided eye you probably won't see the pyramids at Giza, and you certainly won't see the Great Wall of China. Their obscurity is partly the result of having been made from the soil and stone of the surrounding landscape. And although the Great Wall is thousands of miles long, it's only about twenty feet wide—much narrower than the US interstate highways you can barely see from a transcontinental jet.

---

* Adapted from "Exoplanet Earth," *Natural History*, February 2006.

Space Tweet #2
If Earth were size of a school-room globe, you'd find Shuttle and Space Station orbiting 3/8th of an inch above its surface
Apr 19, 2010 5:53 AM

Indeed, from Earth orbit—apart from the smoke plumes rising from the oil-field fires in Kuwait at the end of the first Gulf War in 1991, and the green-brown borders between swaths of irrigated and arid land—the unaided eye cannot see much else that's made by humans. Plenty of natural scenery is visible, though: hurricanes in the Gulf of Mexico, ice floes in the North Atlantic, volcanic eruptions wherever they occur.

From the Moon, a quarter-million miles away, New York, Paris, and the rest of Earth's urban glitter don't even show up as a twinkle. But from your lunar vantage you can still watch major weather fronts move across the planet. Viewed from Mars at its closest, some thirty-five million miles away, massive snow-capped mountain chains and the edges of Earth's continents would be visible through a good backyard telescope. Travel out to Neptune, 2.7 billion miles away—just down the block on a cosmic scale—and the Sun itself becomes embarrassingly dim, now occupying a thousandth the area on the daytime sky that it occupies when seen from Earth. And what of Earth itself? A speck no brighter than a dim star, all but lost in the glare of the Sun.

A celebrated photograph taken in 1990 from the edge of the solar system by the Voyager 1 spacecraft shows how underwhelming Earth looks from deep space: a "pale blue dot," as the American astronomer Carl Sagan called it. And that's generous. Without the help of a picture caption, you might not find it at all.

What would happen if some big-brained aliens from the great beyond scanned the skies with their naturally superb visual organs, further aided by alien state-of-the-art optical accessories? What visible features of planet Earth might they detect?

Blueness would be first and foremost. Water covers more than two-thirds of Earth's surface; the Pacific Ocean alone makes up an entire side of the planet. Any beings with enough equipment and expertise to detect our planet's color would surely infer the presence of water, the third most abundant molecule in the universe.

If the resolution of their equipment were high enough, the aliens would see more than just a pale blue dot. They would see intricate coastlines, too, strongly suggesting that the water is liquid. And smart aliens would surely know that if a planet has liquid water, the planet's temperature and atmospheric pressure fall within a well-determined range.

Earth's distinctive polar ice caps, which grow and shrink from the seasonal temperature variations, could also be seen optically. So could our planet's twenty-four-hour rotation, because recognizable landmasses rotate into view at predictable intervals. The aliens would also see major weather systems come and go; with careful study, they could readily distinguish features related to clouds in the atmosphere from features related to the surface of Earth itself.

Time for a reality check: We live within ten light-years of the nearest known exoplanet—that is, a planet orbiting a star other than the Sun. Most catalogued exoplanets lie more than a hundred light-years away. Earth's brightness is less than one-billionth that of the Sun, and our planet's proximity to the Sun would make it extremely hard for anybody to see Earth directly with an optical telescope. So if aliens have found us, they are likely searching in wavelengths other than visible light—or else their engineers are adapting some other strategy altogether.

Maybe they do what our own planet hunters typically do: monitor stars to see if they jiggle at regular intervals. A star's periodic jiggle betrays the existence of an orbiting planet that may otherwise be too dim to see directly. The planet and host star both revolve around their common center of mass. The more massive the planet, the larger the star's orbit around the center of mass must be, and the more apparent the jiggle when you analyze the star's light. Unfortunately for planet-hunting aliens, Earth is puny, and so the Sun barely budges, posing a further challenge to alien engineers.

Radio waves might work, though. Maybe our eavesdropping aliens have something like the Arecibo Observatory in Puerto Rico, home of Earth's largest single-dish radio telescope—which you might have seen in the early location shots of the 1997 movie *Contact*, based on a novel by Carl Sagan.

If they do, and if they tune to the right frequencies, they'll certainly notice Earth, one of the "loudest" radio sources in the sky. Consider everything we've got that generates radio waves: not only radio itself but also broadcast television, mobile phones, microwave ovens, garage-door openers, car-door unlockers, commercial radar, military radar, and communications satellites. We're just blazing—spectacular evidence that something unusual is going on here, because in their natural state, small rocky planets emit hardly any radio waves at all.

So if those alien eavesdroppers turn their own version of a radio telescope in our direction, they might infer that our planet hosts technology. One complication, though: other interpretations are possible. Maybe they wouldn't be able to distinguish Earth's signal from those of the larger planets in our solar system, all of which are sizable sources of radio waves. Maybe they would think we're a new kind of odd, radio-intensive planet. Maybe they wouldn't be able to distinguish Earth's radio emissions from those of the Sun, forcing them to conclude that the Sun is a new kind of odd, radio-intensive star.

Astrophysicists right here on Earth, at the University of Cambridge in England, were similarly stumped back in 1967. While surveying the skies with a radio telescope for any source of strong radio waves, Anthony Hewish and his team discovered something extremely odd: an object pulsing at precise, repeating intervals of slightly more than a second. Jocelyn Bell, a graduate student of Hewish's at the time, was the first to notice it.

Soon Bell's colleagues established that the pulses came from a great distance. The thought that the signal was technological—another culture beaming evidence of its activities across space—was irresistible. As Bell recounted in an after-dinner speech in 1976, "We had no proof that it was an entirely natural radio emission. . . . Here was I trying to get a Ph.D. out of a new technique, and some silly lot of little green men had to choose my aerial and my frequency to communicate with us." Within a few days, however, she discovered other repeating signals coming from other places in our galaxy. Bell and her associates realized they'd discovered a new class of cosmic object—pulsing stars—which they cleverly, and sensibly, called pulsars.

• • •

Turns out, intercepting radio waves isn't the only way to be snoopy. There's also cosmochemistry. The chemical analysis of planetary atmospheres has become a lively field of modern astrophysics. Cosmochemistry depends on spectroscopy—the analysis of light by means of a spectrometer, which breaks up light, rainbow style, into its component colors. By exploiting the tools and tactics of spectroscopists, cosmochemists can infer the presence of life on an exoplanet, regardless of whether that life has sentience, intelligence, or technology.

The method works because every element, every molecule—no matter where it exists in the universe—absorbs, emits, reflects, and scatters light in a unique way. Pass that light through a spectrometer, and you'll find features that can rightly be called chemical fingerprints. The most visible fingerprints are made by the chemicals most excited by the pressure and temperature of their environment. Planetary atmospheres are crammed with such features. And if a planet is teeming with flora and fauna, its atmosphere will be crammed with biomarkers—spectral evidence of life. Whether biogenic (produced by any or all life-forms), anthropogenic (produced by the widespread species *Homo sapiens*), or technogenic (produced only by technology), this rampant evidence will be hard to conceal.

Unless they happen to be born with built-in spectroscopic sensors, space-snooping aliens would need to build a spectrometer to read our fingerprints. But above all, Earth would have to eclipse its host star (or some other light source), permitting light to pass through our atmosphere and continue on to the aliens. That way, the chemicals in Earth's atmosphere could interact with the light, leaving their marks for all to see.

Some molecules—ammonia, carbon dioxide, water—show up everywhere in the universe, whether life is present or not. But others pop up especially in the presence of life itself. Among the biomarkers in Earth's atmosphere are ozone-destroying chlorofluorocarbons from aerosol sprays, vapor from mineral solvents, escaped coolants from refrigerators and air conditioners, and smog from the burning of fossil fuels. No other way to read that list: sure signs of the absence of intelligence. Another readily detected biomarker is Earth's substantial and sustained level of the molecule methane, more than half of which is produced by human-related activities such as fuel-oil production, rice cultivation, sewage, and the burps of domesticated livestock.

And if the aliens track our nighttime side while we orbit our host star, they might notice a surge of sodium from the sodium-vapor streetlights that switch on at dusk. Most telling, however, would be all our free-floating oxygen, which constitutes a full fifth of our atmosphere.

Oxygen—the third most abundant element in the cosmos, after hydrogen and helium—is chemically active, bonding readily with atoms of hydrogen, carbon, nitrogen, silicon, sulfur, iron, and so on. Thus, for oxygen to exist in a steady state, something must be liberating it as fast as it's being consumed. Here on Earth, the liberation is traceable to life. Photosynthesis, carried out by plants and select bacteria, creates free oxygen in the oceans and in the atmosphere. Free oxygen, in turn, enables the existence of oxygen-metabolizing creatures, including us and practically every other creature in the animal kingdom.

We earthlings already know the significance of Earth's distinctive chemical fingerprints. But distant aliens who come upon us will have to interpret their findings and test their assumptions. Must the periodic appearance of sodium be technogenic? Free oxygen is surely biogenic. How about methane? It, too, is chemically unstable, and yes, some of it is anthropogenic. The rest comes from bacteria, cows, permafrost, soils, termites, wetlands, and other living and nonliving agents. In fact, at this very moment, astrobiologists are arguing about the exact origin of trace amounts of methane on Mars and the copious quantities of methane detected on Saturn's moon Titan, where (we presume) cows and termites surely do not dwell.

If the aliens decide that Earth's chemical features are strong evidence for life, maybe they'll wonder if the life is intelligent. Presumably the aliens communicate with one another, and perhaps they'll presume that other intelligent life-forms communicate too. Maybe that's when they'll decide to eavesdrop on Earth with their radio telescopes to see what part of the electromagnetic spectrum its inhabitants have mastered. So, whether the aliens explore with chemistry or with radio waves, they might come to the same conclusion: a planet where there's advanced technology must be populated with intelligent life-forms, who may occupy themselves discovering how the universe works and how to apply its laws for personal or public gain.

Our catalogue of exoplanets is growing apace. After all, the known universe harbors a hundred billion galaxies, each with hundreds of billions of stars.

The search for life drives the search for exoplanets, some of which probably look like Earth—not in detail, of course, but in overall properties. Those are the planets our descendants might want to visit someday, by choice or by necessity. So far, though, nearly all the exoplanets detected by the planet hunters are much larger than Earth. Most are at least as massive as Jupiter, which is more than three hundred times Earth's mass. Nevertheless, as astrophysicists design hardware that can detect smaller and smaller jiggles of a host star, the ability to find punier and punier planets will grow.

In spite of our impressive tally, planet hunting by earthlings is still in its horse-and-buggy stage, and only the most basic questions can be answered: Is the thing a planet? How massive is it? How long does it take to orbit its host star? No one knows for sure what all those exoplanets are made of, and only a few of them eclipse their host stars, permitting cosmochemists to peek at their atmospheres.

But abstract measurements of chemical properties do not feed the imagination of either poets or scientists. Only through images that capture surface detail do our minds transform exoplanets into "worlds." Those orbs must occupy more than just a few pixels in the family portrait to qualify, and a Web surfer should not need a caption to find the planet in the photo. We have to do better than the pale blue dot.

Only then will we be able to conjure what a faraway planet looks like when seen from the edge of its own star system—or perhaps from the planet's surface itself. For that, we will need spaceborne telescopes with stupendous light-gathering power.

Nope. We're not there yet. But perhaps the aliens are.

••• CHAPTER THREE

EXTRATERRESTRIAL LIFE*

The first half-dozen or so confirmed discoveries of planets around stars other than the Sun—dating to the late 1980s and early 1990s—triggered tremendous public interest. Attention was generated not so much by the discovery of exoplanets but by the prospect of their hosting intelligent life. In any case, the media frenzy that followed was somewhat out of proportion to the events.

Why? Because planets cannot be all that rare in the universe if the Sun happens to have eight of them. Also, the first round of newly discovered planets were all oversize gas giants that resemble Jupiter, which means they have no convenient surface upon which life as we know it could exist. And even if the planets were teeming with buoyant aliens, the odds against these life-forms being intelligent are astronomical.

Ordinarily, there is no riskier step that a scientist (or anyone) can take than to make a sweeping generalization from just one example. At the moment, life on Earth is the only known life in the universe, but compelling arguments suggest that we are not alone. Indeed, nearly all astrophysicists accept the high probability of life elsewhere. The reasoning is easy: if our solar system is not unusual, then the number of planets in the universe would, for example, outnumber the sum of all sounds and words ever

* Adapted from "Is Anybody (Like Us) Out There?" *Natural History*, September 1996, and from "The Search for Life in the Universe: An Overview of the Scientific and Cultural Implications of Finding Life in the Cosmos," congressional testimony presented before the House Committee on Science, Subcommittee on Space and Aeronautics, July 12, 2001, Washington, DC.

uttered by every human who has ever lived. To declare that Earth must be the only planet in the universe with life would be inexcusably big-headed of us.

Many generations of thinkers, both religious and scientific, have been led astray by anthropocentric assumptions and simple ignorance. In the absence of dogma and data, it is safer to be guided by the notion that we are not special, which is generally known as the Copernican principle. It was the Polish astronomer Nicolaus Copernicus who, in the mid-1500s, put the Sun back in the middle of our solar system where it belongs. In spite of a third-century B.C. account of a Sun-centered universe (proposed by the Greek philosopher Aristarchus), the Earth-centered universe has been by far the most popular view for most of the past two thousand years. In the West, it was codified by the teachings of Aristotle and Ptolemy and later by the preachings of the Roman Catholic Church. That Earth was the center of all motion was self-evident: it not only looked that way, but God surely made it so.

The Copernican principle comes with no guarantees that it will guide us correctly for all scientific discoveries yet to come. But it has revealed itself in our humble realization that Earth is not in the center of the solar system, the solar system is not in the center of the Milky Way galaxy, and the Milky Way galaxy is not in the center of the universe. And in case you are one of those people who think that the edge may be a special place, we are not at the edge of anything either.

A wise contemporary posture would be to assume that life on Earth is not immune to the Copernican principle. How, then, can the appearance or the chemistry of life on Earth provide clues to what life might be like elsewhere in the universe?

I do not know whether biologists walk around every day awestruck by the diversity of life. I certainly do. On our planet, there coexist (among countless other life-forms) algae, beetles, sponges, jellyfish, snakes, condors, and giant sequoias. Imagine these seven living organisms lined up next to one another in size-place. If you didn't know better, you would be hard pressed to believe that they all came from the same universe, much less the same planet. And by the way, try describing a snake to somebody who has never seen one: "You gotta believe me! There's this animal on Earth that (1) can stalk its prey with infrared detectors, (2) can swallow whole, live animals several times bigger than its head, (3) has no arms or legs or any other appendage, and yet (4) can travel along the ground at a speed of two feet per second!"

Nearly every Hollywood space movie includes some encounter between humans and alien life-forms, whether from Mars or an unknown planet in a faraway galaxy. The astrophysics in these films serves as the ladder to what people really care about: whether we are alone in the universe. If the person seated next to me on a long airplane flight finds out I'm an astrophysicist, nine times out of ten she'll query me about life in the universe. I know of no other discipline that triggers such consistent enthusiasm from the public.

Given the diversity of life on Earth, one might expect diversity among Hollywood aliens. But I am consistently amazed by the film industry's lack of creativity. With a few notable exceptions—such as the life-forms in *The Blob* (1958) and *2001: A Space Odyssey* (1968)—Hollywood's aliens look remarkably humanoid. No matter how ugly (or cute) they are, nearly all of them have two eyes, a nose, a mouth, two ears, a neck, shoulders, arms, hands, fingers, a torso, two legs, two feet—and they can walk. Anatomically, these creatures are practically indistinguishable from humans, yet they are supposed to have come from another planet. If anything is certain, it is that life elsewhere in the universe, intelligent or otherwise, will look at least as exotic to us as some of Earth's own life-forms do.

Space Tweets #3 & #4

Just drove by the huge, 30-ft tall L-A-X letters near the airport – surely visible from orbit. Is LA an alien space port?

Jan 23, 2010 9:06 AM

Last day in LA. Like the big LAX letters at airport, the HOLLYWOOD sign is huge. Visible from space? Must be where aliens land

Jan 28, 2010 2:16 PM

The chemical composition of Earth-based life is primarily derived from a select few ingredients. The elements hydrogen, oxygen, and carbon account for more than 95 percent of the atoms in the human body and in all other known life. Of the three, it is carbon whose chemical structure allows it to bond most readily and strongly with itself and with many other elements in many different ways—which is why we say life on Earth is carbon-based, and why the study of molecules that contain carbon is generally known as "organic" chemistry. Curiously, the study of life elsewhere in the universe is known as exobiology, one of the few disciplines that attempt to function, at least for now, in the complete absence of firsthand data.

Is life chemically special? The Copernican principle suggests that it probably isn't. Aliens need not look like us to resemble us in more fundamental ways. Consider that the four most common elements in the universe are hydrogen, helium, carbon, and oxygen. Helium is inert. So the three most abundant, chemically active ingredients in the cosmos are also the top three ingredients of life on Earth. For this reason, you can bet that if life is found on another planet, it will be made of a similar mix of elements. Conversely, if life on Earth were composed primarily of manganese and molybdenum, then we would have excellent reason to suspect we're something special in the universe.

Appealing once again to the Copernican principle, we can assume that an alien organism is not likely to be ridiculously large compared with life as we know it. There are cogent structural reasons why you would not expect to find a life-form the size of the Empire State Building strutting around a planet. Even if we ignore the engineering limitations of biological matter, we approach another, more fundamental limit. If we assume that an alien has control of its own appendages, or more generally, if we assume the organism functions coherently as a system, then its size would ultimately be constrained by its ability to send signals within itself at the speed of light—the fastest allowable speed in the universe. For an admittedly extreme example, if an organism were as big as the orbit of Neptune (about ten light-hours across), and if it wanted to scratch its head, then this simple act would take no less than ten hours to accomplish. Subslothlike behavior such as this would be evolutionarily self-limiting, because the time since the beginning of the universe might well be insufficient for the creature to have evolved from smaller forms.

How about intelligence? When Hollywood aliens manage to visit Earth, one might expect them to be remarkably smart. But I know of some that should have been embarrassed by their stupidity. Surfing the FM dial during a car trip from Boston to New York City some years ago, I came upon a radio play in progress that, as best as I could determine, was about evil aliens that were terrorizing earthlings. Apparently, they needed hydrogen atoms to survive, so they kept swooping down to Earth to suck up its oceans and extract the hydrogen from all the H_2O molecules. Now those were some dumb aliens. They must not have been looking at other planets en route to Earth, because Jupiter, for example, contains more than two hundred times the entire mass of Earth in pure hydrogen. I guess nobody told them that more than 90 percent of all atoms in the universe are hydrogen.

And what about aliens that manage to traverse thousands of light-years through interstellar space yet bungle their arrival by crash-landing on Earth?

Then there are the aliens in the 1977 film *Close Encounters of the Third Kind,* who, in advance of their arrival, beam to Earth a mysterious sequence of numbers that is eventually decoded by earthlings to be the latitude and longitude of their upcoming landing site. But Earth's longitude has a completely arbitrary starting point—the prime meridian—which passes through Greenwich, England, by international agreement. And both longitude and latitude are measured in unnatural units we call degrees, 360 of which are in a circle. It seems to me that, armed with this much knowledge of human culture, the aliens could have just learned English and beamed the message "We're going to land a little bit to the side of Devil's Tower National Monument in Wyoming. And because we're arriving in a flying saucer, we won't need runway lights."

Space Tweet #5

Why do aliens always disembark via ramp? Do they have problems with stairs? Or are flying saucers just handicap-accessible?

Aug 21, 2010 12:00 PM

The award for dumbest movie alien of all time must go to the entity that called itself V'ger, from the 1983 film *Star Trek: The Motion Picture*. An ancient mechanical space probe, V'ger had been rescued by a civilization of mechanical aliens and reconfigured so that it could accomplish its mission of discovery across the entire cosmos. The thing grew and grew, acquiring all knowledge of the universe and eventually achieving consciousness. In the film, the crew of the starship Enterprise come upon this now-immense heap of cosmic information and artifacts at a time when V'ger has been searching for its creator. Clued in by the badly tarnished letters "oya" on the original probe, Captain Kirk realizes that V'ger is actually Voyager 6, launched by earthlings in the late twentieth century. Okay. What irks me is how V'ger acquired total knowledge of the cosmos yet remained clueless that its real name was Voyager.

And don't get me started on the 1996 blockbuster *Independence Day*. Actually, I find nothing particularly offensive about evil aliens. There would be no science-fiction film industry without them. The aliens in *Independence Day* are definitely evil. They look like a genetic cross between a Portuguese man-of-war, a hammerhead shark, and a human being. But while they're more creatively conceived than most Hollywood aliens, why are their flying saucers equipped with upholstered high-back chairs with armrests?

I'm glad that, in the end, the humans win. We conquer the *Independence Day* aliens by having a Macintosh laptop computer upload a software virus to the mothership (which happens to be one-fifth the mass of the Moon), thus disarming its protective force field. I don't know about you, but back in 1996 I had trouble just uploading files to other computers within my own department, especially when the operating systems were different. There is only one solution: the entire defense system for the alien mothership must have been powered by the same release of Apple Computer's system software as the laptop computer that delivered the virus.

Let us assume, for the sake of argument, that humans are the only species on Earth to have evolved high-level intelligence. (I mean no disrespect to other big-brained mammals. While most of them cannot do astrophysics, my conclusions are not substantially altered if you wish to include them.) If life on Earth offers any measure of life elsewhere in the universe, then intelligence must be rare. By some estimates, there have been more than ten billion species in the history of life on Earth. It follows that, among all extraterrestrial life-forms, we might expect no better than about one in ten billion to be as intelligent as we are—not to mention the odds against the intelligent life having an advanced technology and a desire to communicate through the vast distances of interstellar space.

Space Tweet #6
Worms dont know that humans who pass by are intelligent, so no reason to think humans would know if alien super-race did same
Jun 3, 2010 9:18 PM

On the chance that such a civilization exists, radio waves would be the communication band of choice because of their ability to traverse the galaxy unimpeded by interstellar gas and dust clouds. But we humans have had command of the electromagnetic spectrum for less than a century. To put that more depressingly: had aliens been trying to send radio signals to earthlings for most of human history, we would have been incapable of receiving them. For all we know, the aliens may have tried to get in touch centuries ago and have concluded that there is no intelligent life on Earth. They would now be looking elsewhere. A more humbling possibility is that aliens did become aware of the technologically proficient species that now inhabits Earth, and drew the same conclusion.

Our Copernican perspective regarding life on Earth, intelligent or otherwise, requires us to presume that liquid water is a prerequisite to life elsewhere. To support life, a planet cannot orbit its host star too closely, or else the temperature would be too high and the planet's water content would vaporize. Also, the orbit should not be too far away, or else the temperature would be too low and the planet's water content would freeze. In other words, conditions on the planet must allow the temperature to stay within the 180°F range of liquid water. As in the three-bowls-of-food scene in "Goldilocks and the Three Bears," the temperature has to be just right. (Once when I was interviewed about this subject on a syndicated radio talk show, the host commented, "Clearly, what you should be looking for is a planet made of porridge!")

While distance from the host planet is an important factor for the existence of life as we know it, a planet's ability to trap stellar radiation matters too. Venus is a textbook example of this "greenhouse" phenomenon. Any visible sunlight that manages to pass through its thick atmosphere of carbon dioxide gets absorbed by Venus's surface and then reradiated in the infrared part of the spectrum. The infrared, in turn, gets trapped by the atmosphere. The unpleasant consequence is an air temperature that hovers at about 900°F, which is much hotter than we would expect, given Venus's distance from the Sun. At that temperature, lead would swiftly become molten.

The discovery of simple, unintelligent life-forms elsewhere in the universe (or evidence that they once existed) would be far more likely—and, for me, only slightly less exciting—than the discovery of intelligent life. Two excellent nearby places to look are beneath the dried riverbeds of Mars (where there may be fossil evidence of life that thrived when waters formerly flowed) and the subsurface oceans that are theorized to exist under the frozen ice layers of Jupiter's moon Europa, whose interior is kept warm by gravitational stresses from the Jovian system. Once again, the promise of liquid water leads our search.

Other common prerequisites for the evolution of life in the universe involve a planet in a stable, nearly circular orbit around a single star. With binary and multiple star systems, which make up more than half of all stars in the galaxy, orbits tend to be strongly elongated and chaotic, which induces extreme temperature swings that would undermine the evolution of stable life-forms. We also require sufficient time for evolution to run its course. High-mass stars are so short-lived (a few million years) that life on Earthlike planets in orbit around them would never have a chance to evolve.

The set of conditions needed to support life as we know it is loosely quantified through what's known as the Drake equation, named for the American astronomer Frank Drake. The Drake equation is more accurately viewed as a fertile idea rather than a rigorous statement of how the physical universe works. It separates the overall probability of finding life in the galaxy into a set of simpler probabilities that correspond to our preconceived notions of suitable cosmic conditions. In the end, after you argue with your colleagues about the value of each probability term in the equation, you are left with an estimate for the total number of intelligent, technologically proficient civilizations in the galaxy. Depending on your bias level—and your knowledge of biology, chemistry, celestial mechanics, and astrophysics—your estimate may range from at least one (ours) up to millions of civilizations in the Milky Way alone.

If we consider the possibility that we may rank as primitive among the universe's technologically competent life-forms—however rare they may be—then the best we can do is to keep alert for signals sent by others, because it is far more expensive to send than to receive. Presumably, an advanced civilization would have easy access to an abundant source of energy, such as its host star. These are the civilizations that would be more likely to do the sending.

The search for extraterrestrial intelligence (affectionately known by its acronym, SETI) has taken many forms. Long-established efforts have relied on monitoring billions of radio channels in search of a radio or microwave signal that might rise above the cosmic noise. The SETI@home screensaver—downloaded by millions of people around the world—enabled a home computer to analyze small chunks of the huge quantities of data collected by the radio telescope at Arecibo Observatory, Puerto Rico. This gigantic "distributed computing" project (the largest in the world) actively tapped the computing power of Internet-connected PCs that would otherwise have been doing nothing while their owners went to the bathroom. More recently, improvements in laser technology have made it worthwhile to search the optical part of the electromagnetic spectrum for pulses of laser light a few nanoseconds in duration. During those nanoseconds, an intense, directed beam of visible light can outshine the light of nearby stars, allowing it to be detected from afar. Another new approach, inspired by the optical version of SETI, is to keep a lookout across the galaxy, not for sustained signals, but for brief blasts of microwaves, which would be relatively cost-efficient to produce on the other end.

The discovery of extraterrestrial intelligence, if and when it happens, will impart a change in human self-perception that may be impossible to anticipate. My only hope is that every other civilization isn't doing exactly what we are doing—because then everybody would be listening, nobody would be sending, and we would collectively conclude there is no other intelligent life in the universe.

Even if we don't soon find life, we will surely keep looking, because we are intellectual nomads—curious beings who derive almost as much fulfillment from the search as we do from the discovery.

••• CHAPTER FOUR

EVIL ALIENS*

Interview with Sanjay Gupta, CNN

Sanjay Gupta: Here's a question: Do you believe in UFOs? If so, you're in some pretty impressive company. British astrophysicist Stephen Hawking, arguably one of the smartest people on the planet, thinks there's a good chance that alien life exists—and not exactly the friendly ET kind. In fact, Hawking envisions a far darker possibility, more along the lines of the movie *War of the Worlds*. In a documentary for the Discovery Channel, Hawking says the aliens will be big, bad, and very busy conquering planet after planet. He says they might live in massive ships, and he calls them nomads who travel the universe conquering others and collecting energy through mirrors. Mirrors; massive ships; giant, mean aliens: is it all possible? Let's go up close with Neil deGrasse Tyson, director of the Hayden Planetarium in New York and, like Hawking, an astrophysicist.

I've been fascinated by this since I was a kid, given the fact that there are hundreds of billions of galaxies, with hundreds of millions of stars in each galaxy.

Neil deGrasse Tyson: Hundreds of *billions* in each galaxy.

SG: Hundreds of billions of stars—even more. And that probably means there's life out there somewhere.

NDT: Indeed.

---

* Adapted from interview with Sanjay Gupta, *Anderson Cooper 360°*, CNN, April 26, 2010.

SG: But this idea that aliens will be evil—Hawking paints a picture that is far less ET and far more *Independence Day*—is this speculation?

NDT: Yes, but it's not blind speculation. It says more about what we fear about ourselves than any real expectations of what an alien would be like. In other words, I think our biggest fear is that the aliens who visit us would treat us the way we treat each other here on Earth. So, in a way, Hawking's apocalyptic fear stories are a mirror held back up to us.

SG: That's a very different perspective than what Carl Sagan put out there. He was literally giving away Earth's location.

NDT: Exactly. Sagan provided the return address on a plaque on the Voyager spacecraft. He wanted to say, "Here's where we are!"

SG: So why would aliens do what Hawking proposes they'll do? Some sort of vengeance?

NDT: Like I said, no one knows how aliens will behave. They will have different chemistry, different motives, different intentions. How can we extrapolate from ourselves to them? Any suspicion that they will be evil is more a reflection of our fear about how *we* would treat an alien species if we found them than any actual knowledge about how an alien species would treat *us*.

Space Tweet #7
How to shield sneezes in space, you ask? Helmet blocks all 40,000 spewed mucous droplets. So Aliens are safe
Jan 15, 2011 2:57 PM

SG: We're listening for them right now. My understanding is that we've been listening for a long time—for anything—and we haven't heard a peep from out there. Do you think they're listening to us right now?

NDT: Possibly. The big fear, it seems to me, is that we announce our presence and then the aliens come and enslave us or put us in a zoo. Some entertaining science-fiction stories have captured just those themes.

SG: I never thought to imagine us as living in an alien zoo.

NDT: That's the fear factor. But what are *we* doing? We're mostly listening. We have giant radio telescopes pointing in different directions, with highly sophisticated circuitry that listens to billions of radio frequencies simultaneously to see if anybody is whispering on any one of them anyplace in the universe. That's different from sending signals out. We're not sending signals out on purpose; we're sending them out accidentally. The expanding edge of our radio bubble is about seventy light-years away right now, and on that frontier you'll find broadcast television shows like *I Love Lucy* and *The Honeymooners*—the first emissaries of human culture that the aliens would decode. Not much reason there for aliens to fear us, but plenty of reason for them to question our intelligence. And, rumors to the contrary, we have not yet heard from aliens, even accidentally. So we're confronting a vacuum, ready to be filled with the many fears we harbor.

••• CHAPTER FIVE

KILLER ASTEROIDS*

The chances that your tombstone will read "KILLED BY ASTEROID" are about the same as they'd be for "KILLED IN AIRPLANE CRASH." Only about two dozen people have been killed by falling asteroids in the past four hundred years, while thousands have died in crashes during the relatively brief history of passenger air travel. So how can this comparative statistic be true? Simple.

The impact record shows that by the end of ten million years, when the sum of all airplane crashes has killed a billion people (assuming a death-by-airplane rate of a hundred per year), an asteroid large enough to kill the same number of people will have hit Earth. The difference is that while airplanes are continually killing people a few at a time, that asteroid might not kill anybody for millions of years. But when it does hit, it will take out a billion people: some instantaneously, and the rest in the wake of global climatic upheaval.

The combined impact rate for asteroids and comets in the early solar system was frighteningly high. Theories of planet formation show that chemically rich gas cooled and condensed to form molecules, then particles of dust, then rocks and ice. Thereafter, it was a shooting gallery. Collisions served as a means for chemical and gravitational forces to bind smaller objects into larger ones. Those objects that, by chance, had accreted slightly more mass than average had slightly higher gravity, attracting other objects

---

* Adapted from "Coming Attractions," *Natural History*, September 1997.

even more. As accretion continued, gravity eventually shaped blobs into spheres, and planets were born. The most massive planets had sufficient gravity to retain the gaseous envelope we call an atmosphere.

Every planet continues to accrete, every day of its life, although at a significantly lower rate than when it first formed. Even today, interplanetary dust rains down on Earth in vast quantities—typically a hundred tons of it a day—though only a small fraction reaches Earth's surface. The rest harmlessly vaporizes in Earth's atmosphere as shooting stars. More hazardous are the billions, likely trillions, of leftover rocks—comets and asteroids—that have been orbiting the Sun since the early years of our solar system but haven't yet managed to join up with a larger object.

Long-period comets—icy vagabonds from the extreme reaches of the solar system (as much as a thousand times the radius of Neptune's orbit)—are susceptible to gravitational nudges from passing stars and interstellar clouds, which can direct them on a long journey inward toward the Sun, and therefore to our neighborhood. Several dozen short-period comets from the nearer reaches of the solar system are known to cross Earth's orbit.

As for the asteroids, most are made of rock. The rest are metal, mostly iron. Some are rubble piles—gravitationally bound collections of bits and pieces. Most asteroids live between the orbits of Mars and Jupiter and will never ever come near Earth.

But some do. Some will. About ten thousand near-Earth asteroids are known, with more surely to be discovered. The most threatening of them number more than a thousand, and that number steadily grows as spacewatchers continually survey the skies in search of them. These are the "potentially hazardous asteroids," all larger than about five hundred feet across, with orbits that bring them within about twenty times the distance between Earth and the Moon. Nobody's saying they're all going to hit tomorrow. But all of them are worth watching, because a little cosmic nudge here or there might just send them a little closer to us.

In this game of gravity, by far the scariest impactors are the long-period comets—those whose orbits around the Sun take longer than two hundred years. Representing about one-fourth of Earth's total risk of impacts, such comets fall toward the inner solar system from gargantuan distances and achieve speeds in excess of a hundred thousand miles an hour by the time they reach Earth. Long-period comets thus achieve more awesome impact energy for their size than your run-of-the-mill asteroid. More important, they are too distant, and too dim, throughout most of their orbit to be reliably tracked. By the time a long-period comet is discovered to be heading our way, we might have just a couple of years—or a couple of months—to fund, design, build, and launch a craft to intercept it. In 1996, for instance, comet Hyakutake was discovered only four months before its closest approach to the Sun because its orbit was tipped strongly out of the plane of our solar system, precisely where nobody was looking. While en route, it came within ten million miles of Earth: a narrow miss.

The term "accretion" is duller than "species-killing, ecosystem-destroying impact," but from the point of view of solar-system history, the terms are the same. Impacts made us what we are today. So, we cannot simultaneously be happy that we live on a planet, happy that our planet is chemically rich, and happy that dinosaurs don't rule the Earth, and yet resent the risk of a planet-wide catastrophe.

In a collision with Earth, some of an impactor's energy gets deposited into our atmosphere through friction and an airburst of shock waves. Sonic booms are shock waves too, but they're typically made by airplanes with speeds between one and three times the speed of sound. The worst damage they might do is jiggle the dishes in your china cabinet. But at speeds in excess of 45,000 miles per hour—nearly seventy times the speed of sound—the shock waves from the average collision between an asteroid and Earth can be devastating.

If the asteroid or comet is large enough to survive its own shock waves, the rest of its energy gets deposited on Earth. The impact blows a crater up to twenty times the diameter of the original object and melts the ground below. If many impactors hit one after another, with little time between each strike, then Earth's surface will not have enough time to cool between impacts. We infer from the pristine cratering record on the surface of our nearest neighbor, the Moon, that Earth experienced such an era of heavy bombardment between 4.6 billion and 4.0 billion years ago.

The oldest fossil evidence for life on Earth dates from about 3.8 billion years ago. Before that, Earth's surface was being relentlessly sterilized. The formation of complex molecules, and thus life, was inhibited, although all the basic ingredients were being delivered. That would mean it took 800 million years for life to emerge here (4.6 billion – 3.8 billion = 800 million). But to be fair to organic chemistry, you must first subtract all the time that Earth's surface was forbiddingly hot. That leaves a mere 200 million years for life's emergence from a rich chemical soup—which, like all good soups, included liquid water.

Much of that water was delivered to Earth by comets more than four billion years ago. But not all space debris is left over from the beginning of the solar system. Earth has been hit at least a dozen times by rocks ejected from Mars, and we've been hit countless more times by rocks ejected from the Moon.

Ejections occur when impactors carry so much energy that, when they hit, smaller rocks near the impact zone are thrust upward with sufficient speed to escape a planet's gravitational grip. Afterward, those rocks mind their own ballistic business in orbit around the Sun until they slam into something. The most famous of the Mars rocks is the first meteorite found near the Alan Hills section of Antarctica in 1984—officially known by its coded (though sensible) abbreviation, ALH-84001. This meteorite contains tantalizing, yet circumstantial, evidence that simple life on the Red Planet thrived a billion years ago.

Mars has abundant "geo"-logical evidence—dried river beds, river deltas, floodplains, eroded craters, gullies on steep slopes—for a history of running water. There's also water there today in frozen form (polar ice caps and plenty of subsurface ice) as well as minerals (silica, clay, hematite "blueberries") that form in standing water. Since liquid water is crucial to the survival of life as we know it, the possibility of life on Mars does not stretch scientific credulity. The fun part comes with the speculation that life-forms first arose on Mars and were blasted off the planet's surface, thus becoming the solar system's first microbial astronauts, arriving on Earth to jump-start evolution. There's even a word for that process: panspermia. Maybe we are all Martians.

Matter is far more likely to travel from Mars to Earth than vice versa.

Escaping Earth's gravity requires more than two and a half times the energy required to leave Mars. And since Earth's atmosphere is about a hundred times denser, air resistance on Earth (relative to Mars) is formidable. Bacteria on a voyaging asteroid would have to be hardy indeed to survive several million years of interplanetary wanderings before plunging to Earth. Fortunately, there is no shortage of liquid water and rich chemistry here at home, so even though we still cannot definitively explain the origin of life, we do not require theories of panspermia to do so.

Of course, it's easy to think impacts are bad for life. We can and do blame them for major episodes of extinction in the fossil record. That record displays no end of extinct life-forms that thrived far longer than the current Earth tenure of *Homo sapiens*. Dinosaurs are among them. But what are the ongoing risks to life and society?

House-size impactors collide with Earth, on average, every few decades. Typically they explode in the atmosphere, leaving no trace of a crater. But even baby impacts could become political time bombs. If such an atmospheric explosion occurred over India or Pakistan during one of the many episodes of escalated tension between those two nations, the risk is high that someone would misinterpret the event as a first nuclear strike, and respond accordingly. At the other end of the impactor scale, once in about a hundred million years we're visited by an impactor capable of annihilating all life-forms bigger than a carry-on suitcase. In cases such as those, no political response would be necessary.

Space Tweet #8

For some people, space is irrelevant. But when the asteroid comes, I bet they'll think differently

Apr 13, 2011 8:40 PM

What follows is a table that relates average collision rates on Earth to the size of the impactor and the equivalent energy in millions of tons of TNT. It's based on a detailed analysis of the history of impact craters on Earth, the erosion-free cratering record on the Moon's surface, and the known numbers of asteroids and comets whose orbits cross that of Earth. These data are adapted from a congressionally mandated study titled *The Spaceguard Survey: Report of the NASA International Near-Earth Object Detection Workshop.* For comparison, the table includes the impact energy in units of the atomic bomb dropped by the US Air Force on Hiroshima in 1945.

••• RISK OF IMPACTS ON EARTH •••

Once per	Asteroid Diameter (meters)	Impact Energy (megatons of TNT)	Impact Energy (atomic bomb equivalent)
Month	3	0.001	0.05
Year	6	0.01	0.5
Decade	15	0.2	10
Century	30	2	100
Millennium	100	50	2,500
10,000 years	200	1,000	50,000
1,000,000 years	2,000	1,000,000	50,000,000
100,000,000 years	10,000	100,000,000	5,000,000,000

The energetics of some famous impacts can be located on the table. For example, a 1908 explosion near the Tunguska River in Siberia felled thousands of square kilometers of trees and incinerated the three hundred square kilometers that encircled ground zero. The culprit is believed to have been a sixty-meter stony meteorite (about the size of a twenty-story building) that exploded in midair, thus leaving no crater. The chart indicates that collisions of this magnitude happen, on average, every couple of centuries. A much rarer sort of event created the nearly two-hundred-kilometer-wide Chicxulub crater on Mexico's Yucatán Peninsula, which is believed to have been left by an asteroid perhaps ten kilometers wide, with an impact energy five billion times greater than the atomic bombs exploded in World War II. This is one of those collisions that take place once in a hundred million years. The crater dates from about sixty-five million years ago, and there hasn't been one of similar magnitude since.

Coincidentally, at about the same time, *Tyrannosaurus rex* and friends became extinct, enabling mammals to evolve into something more ambitious than tree shrews.

It's useful to consider how strikes by comets and asteroids impact Earth's ecosystem. In a fat book titled *Hazards Due to Comets and Asteroids*, several planetary scientists do just that regarding these unwelcome deposits of energy. Here's a bit of what they sketched out:

- Most impactors with less than about ten megatons of energy will explode in the atmosphere, leaving no trace of a crater. The few that survive in one piece are likely to be iron based.

- A blast of 10 to 100 megatons from an iron asteroid will make a crater, whereas its stony equivalent will disintegrate, producing primarily airbursts. On land, the iron impactor will destroy an area equivalent to Washington, DC.

- A land impact of 1,000 to 10,000 megatons will produce a crater and destroy an area the size of Delaware. An oceanic impact of that magnitude will produce significant tidal waves.

- A blast of 100,000 to 1,000,000 megatons will result in global destruction of ozone. An oceanic impact will generate tidal waves on an entire hemisphere, while a land impact will raise enough dust into the stratosphere to alter Earth's weather and freeze crops. A land impact will destroy an area the size of France.

- A blast of 10,000,000 to 100,000,000 megatons will result in prolonged climatic change and global conflagration. A land impact will destroy an area equivalent to the continental United States.

- A blast of 100,000,000 to 1,000,000,000 megatons, whether on land or sea, will lead to mass extinction on the scale of the Chicxulub impact, when three-quarters of Earth's species were wiped out.

Earth, of course, is not the only rocky planet at risk of impacts. Mercury has a cratered face that, to a casual observer, looks just like the Moon. Radio topography of cloud-enshrouded Venus shows no shortage of craters. And Mars, with its historically active geology, reveals large, recently formed craters.

At more than three hundred times the mass of Earth, and more than ten times its diameter, Jupiter's ability to attract impactors is unmatched among the planets of our solar system. In 1994, during the week of anniversary celebrations for the twenty-fifth anniversary of the Apollo 11 Moon landing, comet Shoemaker-Levy 9, having broken into a couple dozen chunks during a previous close encounter with Jupiter, slammed—one chunk after another, at a speed of more than 200,000 kilometers an hour—into the Jovian atmosphere. Backyard telescopes down here on Earth easily detected the gaseous scars. Because Jupiter rotates swiftly (once every ten hours), each piece of the comet plunged into a different location as the atmosphere slid by.

In case you were wondering, each piece of Shoemaker-Levy 9 hit with the equivalent energy of the Chicxulub impact. So, whatever else is true about Jupiter, it surely has no dinosaurs left.

You'll be happy to learn that in recent years, more and more planetary scientists around the world have gone in search of vagabonds from space that might be heading our way. True, our list of potential killer impactors is incomplete, and our ability to predict the behavior of objects millions of orbits into the future is severely compromised by the onset of chaos. But we can focus on what will happen in the next few decades or centuries.

Among the population of Earth-crossing asteroids, we have a chance at cataloguing everything larger than about one kilometer wide—the size that begins to wreak global catastrophe. An early-warning and defense system to protect the human species from these impactors is a reachable goal. Unfortunately, objects much smaller than a kilometer, of which there are many, reflect much less light and are therefore much harder to detect and track. Because of their dimness, they can hit us without notice—or with notice far too short for us to do anything about them. In January 2002, for instance, a stadium-size asteroid passed by at about twice the distance from here to the Moon—and it was discovered just twelve days before its closest approach. Given another decade or so of data collecting and detector improvements, however, it may be possible to catalogue nearly all asteroids down to about 140 meters across. While the small stuff carries enough energy to incinerate entire nations, it will not put the human species at risk of extinction.

Any of these we should worry about? At least one. On Friday the 13th, April 2029, an asteroid large enough to fill the Rose Bowl as though it were an egg cup will fly so close to Earth that it will dip below the altitude of our communication satellites. We did not name this asteroid Bambi. Instead, we named it Apophis, after the Egyptian god of darkness and death. If the trajectory of Apophis at close approach passes within a narrow range of altitudes called the "keyhole," then the influence of Earth's gravity on its orbit will guarantee that seven years later, in 2036, on its next trip around, the asteroid will hit Earth directly, likely slamming into the Pacific Ocean between California and Hawaii. The five-story tsunami it creates will wipe out the entire west coast of North America, dunk Hawaiian cities, and devastate all the landmasses of the Pacific Rim. If Apophis misses the keyhole in 2029, we will have nothing to worry about in 2036.

Once we mark our calendars for 2029, we can either pass the time sipping cocktails at the beach and planning to hide from the impact, or we can be proactive.

The battle cry of those anxious to wage nuclear war is "Blow it out of the sky!" True, the most efficient package of destructive energy ever conceived by humans is nuclear power. A direct hit on an incoming asteroid might explode it into enough small pieces to reduce the impact danger to a harmless, though spectacular, meteor shower. Note that in empty space, where there is no air, there can be no shock waves, and so a nuclear warhead must actually make contact with the asteroid to do damage.

Another method would be to engage a radiation-intensive neutron bomb (that's the Cold War–era bomb that kills people but leaves buildings intact). The bomb's high-energy neutron bath would heat up one side of the asteroid, causing material to spew forth and thus induce the asteroid to recoil. That recoil would alter the asteroid's orbit and remove it from the collision path.

A kindler, gentler method would be to nudge the asteroid out of harm's way with slow but steady rockets that have somehow been attached to one side. Apart from the uncertainty of how to attach rockets to an unfamiliar material, if you do this early enough, then all you need is a small push using conventional chemical fuels. Or maybe you attach a solar sail, which harnesses the pressure of sunlight for its propulsion, in which case you'll need no fuel at all.

The odds-on favorite solution, however, is the gravitational tractor. This involves parking a probe in space near the killer asteroid. As their mutual gravity draws the probe to the asteroid, retro rockets fire, instead causing the asteroid to draw toward the probe and off its collision course with Earth.

The business of saving the planet requires commitment. We must first catalogue every object whose orbit intersects Earth's. We must then perform precise computer calculations that enable us to predict a catastrophic collision hundreds or thousands of orbits into the future. Meanwhile, we must also carry out space missions to determine in great detail the structure and chemical composition of killer comets and asteroids. Military strategists understand the need to know your enemy. But now, for the first time, we would be engaged in a space mission conceived not to beat a spacefaring competitor but to protect the life of our entire species on our collective planetary home.

Whichever option we choose, we will first need that detailed inventory of orbits for all objects that pose a risk to life on Earth. The number of people in the world engaged in that search totals a few dozen. I'd feel more comfortable if there were a few more. The decision comes down to how long into the future we're willing to protect the life of our own species on Earth. If humans one day become extinct from a catastrophic collision, it won't be because we lacked the brainpower to protect ourselves, but because we lacked the foresight and determination. The dominant species that replaces us on postapocalyptic Earth might just wonder why we fared no better than the proverbially pea-brained dinosaurs.

••• CHAPTER SIX

DESTINED FOR THE STARS*

Video interview with Calvin Sims for *The New York Times*

The Conversation

Neil deGrasse Tyson: We need to go back to the Moon. Many people say, "We've been there, done that, can't you come up with a new place to visit?" But the Moon offers important technological advantages. A trip to Mars takes about nine months. If you haven't been out of low Earth orbit for forty years, sending people to Mars for the first time is a long way to go and a hard thing to do. A big thrust of the new space vision is to reengage the manned program in ways that haven't been done during the past decade, and to recapture the excitement that drove so much of the space program back in the 1960s.

Calvin Sims: So the reasons to go are to prove that we can do it again, because we haven't done it in such a long time, and also to build consensus for it?

NDT: We haven't left low Earth orbit recently. We have to remind ourselves how to do that—how to do it well, how to do it efficiently. We also have to figure out how to set up base camp and sustain life in a place other than Earth or low Earth orbit. The Moon is a relatively easy place to get to and test all this out.

---

* Adapted from "The Conversation: Neil Tyson," *The New York Times* video, conducted June 23, 2006, by Calvin Sims, produced by Matt Orr; posted online July 20, 2006, at http://www.nytimes.com/ref/multimedia/conversation.html.

CS: NASA has estimated it could cost $100 billion, conservatively, to go to the Moon. Do you think it's prudent to be funding this effort, especially at a point in our history when we have a war in Iraq and a lot of domestic demands?

NDT: This $100 billion figure needs to be unpacked. It doesn't come all at once; it's spread over multiple years. And $100 billion, by the way, is only six years of total NASA funding.

America is a wealthy nation. Let's ask the question, "What is going to space worth to you?" How much of your tax dollar are you willing to spend for the journey that NASA represents in our heart, in our minds, in our souls? NASA's budget comes to one-half of one percent of your tax bill. So I don't think that's the first place people should be looking if they want to save money in the federal budget. It's certainly worth a whole percent—personally, I think it's worth more than that—but if all you're going to give us is one percent, we can make good use of it.

Destined for the Stars

NDT: In every culture across time, there has always been somebody wondering about our place in the universe and trying to come to terms with what Earth is. This is not a latter-day interest; it's something deeply inherent in what it is to be human. As twenty-first-century Americans, we're lucky to be able to act on that wonder. Most people just stood there, looked upward, and invented mythologies to explain what they were wondering about. We actually get to build spaceships and go places. That's a privilege brought by the success of our economy and the vision of our leaders, combined with the urge to do it in the first place.

CS: You're saying the primary reason to venture into space is the quest for knowledge, and that humans are programmed by nature to satisfy our curiosity and to engage in the sheer thrill of discovery. Why is the allure so great that we risk human lives to get there?

NDT: Not everyone would risk their life. But for some members of our species, discovery is fundamental to their character and identity. And those among us who feel that way then carry the nation, the world, into the future.

Robots are important also. If I don my pure-scientist hat, I would say just send robots; I'll stay down here and get the data. But nobody's ever given a parade for a robot. Nobody's ever named a high school after a robot. So when I don my public-educator hat, I have to recognize the elements of exploration that excite people. It's not only the discoveries and the beautiful photos that come down from the heavens; it's the vicarious participation in discovery itself.

CS: How far are we from having mass space exploration and experience by the individual person—the colonization of space? This has been a dream for a long time. Is it twenty years off? Thirty years?

NDT: Anytime I read about the history of human behavior, I see that people are always finding some reason to fight and kill one another. This is really depressing. And so I don't know that I trust human beings to colonize another planet, and to keep those colonies from becoming zones of violence and conflict. Also, the future has been a little oversold. Just look at what people said in the 1960s: "By 1985 there will be thousands of people living and working in space." No. It's now 2006, and we've got three people living and working in space. Delusions come about because people lose track of the forces that got us into space in the first place.

CS: Do you have any desire yourself to venture out and explore space?

NDT: No, never. Part of the popular definition of the word "space" is, for example, to go into Earth orbit. Well, Earth orbit can be as low as two hundred miles above Earth's surface. That's the distance from New York to Boston. My interest in space goes vastly beyond that—to galaxies, black holes, the Big Bang. Now, if we had ways to travel that far, sure, sign me up. Visit the Andromeda galaxy? I'm ready to leave tomorrow. But we don't have a way to do that yet, so I'll sit back and wait for it to come.

The Sun Revolves Around the Earth?

CS: Americans on average know far less about science and technology than their foreign counterparts. You've said that unless we take steps to improve scientific literacy in America, we are headed for a crisis.

NDT: The crisis is happening already. But I'm pleased to report that people with understanding and foresight are in our midst, some of whom have served on committees that produce documents. "A Nation at Risk," the 1983 report by the National Commission on Excellence in Education, for example, commented that if an enemy power tried to impose on America the substandard educational system that exists today, we might have considered it an act of war. In fact, the report went so far as to say that America had essentially been "committing an act of unthinking, unilateral educational disarmament."

CS: Some studies have shown that only about 20 to 25 percent of the adult population can be considered scientifically literate. And one study found that one American adult in five thinks that the Sun revolves around the Earth, a notion that was abandoned in the sixteenth century. Does that surprise you?

NDT: Didn't you just ask me whether we're in a crisis? Yes, we are. And yes, it concerns me deeply. There's fundamental knowledge about the physical world that the general public is oblivious to. And by the way, science literacy is not simply how many chemical formulas you can recite, nor whether you know how your microwave oven works. Science literacy is being plugged into the forces that power the universe. There is no excuse for thinking that the Sun, which is a million times the size of Earth, orbits Earth.

CS: This is particularly troubling because so much political debate has a basis in science: global warming, stem cell research. What do we do about this?

NDT: I can only tell you what *I* do about it. I hate to say this, but I've given up on adults. They've formed their ways; they're the product of whatever happened in their lives; I can't do anything for them. But I can have some influence on people who are still in school. That's where I, as a scientist and an educator, can do something to help teach them how to think, how to evaluate a claim, how to judge what one person says versus what another says, how to establish a level of skepticism. Skepticism is healthy. It's not a bad thing; it's a good thing. So I'm working on the next generation as they come up. I don't know what to do with the rest. That 80 percent of the adults, I can't help you there.

CS: How do we change the way science is taught?

NDT: Ask anybody how many teachers truly made a difference in their life, and you never come up with more than the fingers on one hand. You remember their names, you remember what they did, you remember how they moved in front of the classroom. You know why you remember them? Because they were passionate about the subject. You remember them because they lit a flame within you. They got you excited about a subject you didn't previously care about, because they were excited about it themselves. That's what turns people on to careers in science and engineering and mathematics. That's what we need to promote. Put that in every classroom, and it will change the world.

China: The New Sputnik

NDT: It's sad but true that one of the biggest drivers fueling the space program in the 1960s was the Cold War. We don't remember it that way; instead we remember it as, "We're Americans, and we're explorers." What actually happened was that Sputnik lit a flame under our buns, and we said, "This is not good. The Soviet Union is our enemy, and we have to beat them."

CS: Now China is the competitor. So would you say America's ambitious new space initiative is being driven by economic and military goals, especially since China put a man into orbit in 2003 and is close to reaching the Moon?

NDT: There's a proximity in time between the launch of the first Chinese taikonaut into orbit, which was October 2003, and a spate of US documents articulating a "space vision," including the Bush administration's Vision for Space Exploration in January 2004 and an executive order that same month establishing the Presidential Commission on Implementation of United States Space Exploration Policy, followed by NASA's Vision for Space Exploration in February. The space vision does not state, "We're worried about the Chinese; let's get our people back into orbit," but it would be imprudent not to reflect on the political climate in which these documents were issued. I have no doubt that we're worried about our ability to compete. Let's not forget that the vision was announced within a year of the loss of the Columbia space shuttle. It was in the wake of that loss that people started asking questions: What is NASA doing with its manned program? Why are we risking our lives to just drive around the block, boldly going where hundreds have gone before? If you're going to put your life at risk, let it be because you're going somewhere no one has ever been. It's not about being risk averse: you want the risk to be matched to the goal.

CS: How far advanced are the Chinese? Can we beat them back to the Moon?

NDT: Of the many comparative statistics between America and other nations, one of my favorites is that there are more scientifically literate people in China than there are college graduates here in America. When I was on the president's aerospace commission, we went around the world to study the economic climate that our own aerospace industry was now competing in. One of those trips was to China. We met with government officials and industry leaders in 2002—by the way, they all had rings from engineering schools in America—and they told us, "We're going to put a man in space in a few years." There was no doubt in our minds that this would happen, because we saw the channeling of their resources into this effort. We saw how they valued it for national pride. We saw how they valued it as an economic engine. What's fresh for them is what too many Americans have taken for granted within our own nation.

CS: Is the militarization of space or the colonization of space by different countries inevitable as a consequence of our getting there?

NDT: We've got lots of space assets: communications satellites, weather satellites, GPS. There's talk of protecting those. Is that the militarization of space that people refer to? Maybe instead they're referring to lasers and bombs. If that were the trend, it would not be good. Militarization would contaminate the purity of the vision. The vision is to explore. There's nothing purer in the human spirit than that.

Losing Our Scientific Edge

CS: The United States remains the dominant scientific and technological power in the world, but foreign competitors are gaining ground, are they not?

NDT: It's not that we're losing our edge; it's that everyone's catching up with us. The United States maintained our investments on technological frontiers in the 1950s, '60s, and '70s. We could have stayed ahead of the world, as we were during those decades. Yes, everyone caught up with us and leveled the playing field—but it didn't have to stay that way. And it doesn't have to stay that way now. Time for us to reinvest in ourselves. Our nation has the largest economy in the world; it's not out of our reach to reclaim the leadership we once had.

CS: But fewer and fewer students are majoring in science and engineering, and, in fact, a substantial portion of our scientific and technological workforce is foreign-born. Isn't this a concern?

NDT: I'm not concerned, per se, that foreign students fill a substantial part of our educational pipeline in science and engineering. It's been that way for several decades. America loses only if those students go home.

CS: Is that happening?

NDT: Yes, it is. Before, foreign students would come and stay, and so our investments in them as students produced a return in their creativity and innovation as workers. They became part of the American economy.

CS: So why are they going back home now?

NDT: Because the rest of the world is catching up, and now there are opportunities back in their native countries—opportunities that vastly exceed what's available here.

CS: Isn't the increase, the infusion, of scientific capability good for science? Isn't that what you want to happen?

NDT: It depends what day you catch me and which hat I'm wearing. It's easy to speak in terms of wanting to keep America strong, healthy, and wealthy. But as a scientist, you really only care about the frontier of science, wherever that frontier arises. Yes, you want to be on that frontier yourself, but science has always been international. In some ways science transcends nationality, because all scientists speak the same language. The equations are the same, no matter what side of the ocean you're on or when you've written them. So ultimately, yes, it's good that more people are doing science and that more countries embrace investments in science. Nevertheless, I'll lament the day Americans become bystanders rather than leaders on the space frontier.

CHAPTER SEVEN

WHY EXPLORE*

Unlike other animals, humans are quite comfortable sleeping on our backs. This simple fact affords us a view of the boundless night sky as we fall asleep, allowing us to dream about our place in the cosmos and to wonder what lies undiscovered in the worlds beyond. Or perhaps a gene operates within us that demands we learn for ourselves what awaits us on the other side of the valley, over the seas, or across the vacuum of space. Regardless of the cause, the effect is to leave us restless for want of a plan to discover. We know in our minds, but especially in our hearts, the value to our culture of new voyages and the new vistas they provide. Because without them, our culture stalls and our species withers. And we might as well go to sleep facing down.

---

* Adapted from "Why Explore?" in Lonnie Jones Schorer (with a foreword by Buzz Aldrin), *Kids to Space: A Space Traveler's Guide* (Burlington, Ontario: Collector's Guide Publishing, 2006).

••• CHAPTER EIGHT

THE ANATOMY OF WONDER*

These days we wonder about many things. We wonder whether we will arrive at work on time. We wonder whether the recipe for corn muffins we got off the Internet will turn out okay. We wonder whether we will run out of fuel before reaching the next gas station. As an intransitive verb, *wonder* is just another word in a sentence. But as a noun (with the exception of "Boy Wonder," the moniker for Batman's sidekick), the word expresses one of our highest capacities for human emotion.

Most of us have felt wonder at one time or another. We come upon a place or thing or idea that defies explanation. We behold a level of beauty and majesty that leaves us without words; awe draws us into a state of silent stupor. What's remarkable is not that humans are endowed with this capacity to feel, but that very different forces can stimulate these same emotions within us all.

The reverent musings of a scientist at the boundary of what is known and unknown in the universe—on the brink of cosmic discovery—greatly resembles the thoughts expressed by a person steeped in religious reverence. And (as is surely the goal of most artists) some creative works leave the viewer without words—only feelings that hover at the limits of the emotional spectrum. The encounter is largely spiritual and cannot be absorbed all at once; it requires persistent reflection on its meaning and on our relationship to it.

---

* Adapted from "Wonder," posted at abc.com, October 30, 2006.

Each component of this trinity of human endeavor—science, religion, and art—lays powerful claim to our feelings of wonder, which derive from an embrace of the mysterious. Where mystery is absent, there can be no wonder.

Viewing a great work of engineering or architecture can force one to pause out of respect for the sublime intersection of science and art. Projects of such a scale have the power to transform the human landscape, announcing loudly, both to ourselves and to the universe, that we have mastered the forces of nature that formerly bound us to an itinerant life in search of food, shelter, and nothing else.

Inevitably, new wonders supplant old wonders, induced by modern mysteries instead of old. We must ensure that this forever remains true, lest our culture stagnate through time and space. Two thousand years ago, long before we understood how and why the planets moved the way they do in the night sky, the Alexandrian mathematician and astronomer Claudius Ptolemy could not restrain his reverence as he contemplated them. In the *Almagest* he writes: "When I trace, at my pleasure, the windings to and fro of the heavenly bodies, I no longer touch Earth with my feet. I stand in the presence of Zeus himself and take my fill of ambrosia."

People no longer wax poetic about the orbital paths of planets. Isaac Newton solved that mystery in the seventeenth century with his universal law of gravitation. That Newton's law is now taught in high school physics classes stands as a simple reminder that on the ever-advancing frontier of discovery, on Earth and in the heavens, the wonders of nature and of human creativity know no bounds, forcing us periodically to reassess what to call the most wondrous.

••• CHAPTER NINE

HAPPY BIRTHDAY, NASA*

Dear NASA,

Happy birthday! Perhaps you didn't know, but we're the same age. In the first week of October 1958, you were born of the National Aeronautics and Space Act as a civilian space agency, while I was born of my mother in the East Bronx. So the yearlong celebration of our golden anniversaries, which began the day after we both turned forty-nine, provides me a unique occasion to reflect on our past, present, and future.

I was three years old when John Glenn first orbited Earth. I was eight when you lost astronauts Chaffee, Grissom, and White in that tragic fire of their Apollo 1 capsule on the launchpad. I was ten when you landed Armstrong and Aldrin on the Moon. And I was fourteen when you stopped going to the Moon altogether. Over that time I was excited for you and for America. But the vicarious thrill of the journey, so prevalent in the hearts and minds of others, was absent from my emotions. I was obviously too young to be an astronaut. But I also knew that my skin color was much too dark for you to picture me as part of this epic adventure. Not only that, even though you are a civilian agency, your most celebrated astronauts were military pilots, at a time when war was becoming less and less popular.

During the 1960s, the civil rights movement was surely more real to me than to you. In fact, it took a directive from Vice President Johnson

---

* First published in *NASA 50th Magazine: 50 Years of Exploration and Discovery,* 2008.

in 1963 to force you to hire black engineers at your prestigious Marshall Space Flight Center in Huntsville, Alabama. I found the correspondence in your archives. Do you remember? James Webb, then head of NASA, wrote to German rocket pioneer Wernher von Braun, who headed the center and was the chief engineer of the entire manned space program. The letter boldly and bluntly directs von Braun to address the "lack of equal employment opportunity for Negroes" in the region, and to collaborate with the region's colleges Alabama A&M and Tuskegee to identify, train, and recruit qualified Negro engineers into the NASA Huntsville family.

In 1964, you and I had not yet turned six when I saw picketers outside the newly built apartment complex of our choice, in the Riverdale section of the Bronx. They were protesting to prevent Negro families, mine included, from moving there. I'm glad their efforts failed. These buildings were called, perhaps prophetically, the Skyview Apartments, on whose roof, twenty-two stories above the Bronx, I would later train my telescope on the universe.

My father was active in the civil rights movement, working under New York City's Mayor Lindsay to create job opportunities for youth in the ghetto, as the "inner city" was called back then. Year after year, the forces operating against this effort were huge: poor schools, bad teachers, meager resources, abject racism, and assassinated leaders. So while you were celebrating your monthly advances in space exploration from Mercury to Gemini to Apollo, I was watching America do all it could to marginalize who I was and what I wanted to become in life.

I looked to you for guidance, for a vision statement that I could adopt and that would fuel my ambitions. But you weren't there for me. Of course, I shouldn't blame you for society's woes. Your conduct was a symptom of America's habits, not a cause. I knew this. But you should nonetheless know that among my colleagues, I am the only one in my generation who became an astrophysicist *in spite of* your achievements in space rather than *because of* them. For my inspiration, I instead turned to libraries, remaindered books on the cosmos from bookstores, my rooftop telescope, and the Hayden Planetarium. After some fits and starts through my years in school, when becoming an astrophysicist seemed at times to be the path of most resistance through an unwelcoming society, I became a professional scientist. I became an astrophysicist.

Over the decades that followed, you've come a long way—including, most recently, a presidentially initiated, congressionally endorsed vision statement that finally gets us back out of low Earth orbit. Whoever does not yet recognize the value of this adventure to our nation's future soon will, as the rest of the developed and developing world passes us by in every measure of technological and economic strength. Not only that, today you look much more like America—from your senior-level managers to your most decorated astronauts. Congratulations. You now belong to the entire citizenry. Examples of this abound, but I especially remember in 2004 when the public rallied around the Hubble Telescope, your most beloved unmanned mission. They all spoke loudly, ultimately reversing the threat that the telescope's life might not be extended for another decade. Hubble's transcendent images of the cosmos had spoken to us all, as did the personal profiles of the space shuttle astronauts who deployed and serviced the telescope, and the scientists who benefited from its data stream.

Not only that, I've even joined the ranks of your most trusted, as I served dutifully on your advisory council. I came to recognize that when you're at your best, nothing in this world can inspire the dreams of a nation the way you can—dreams carried by a parade of ambitious students, eager to become scientists, engineers, and technologists in the service of the greatest quest there ever was. You have come to represent a fundamental part of America's identity, not only to itself but to the world.

So, now that we've both turned forty-nine and are well into our fiftieth orbit around the Sun, I want you to know that I feel your pains and share your joys. And I look forward to seeing you back on the Moon. But don't stop there. Mars beckons, as do destinations beyond.

Birthday buddy, even if I have not always been, I am now your humble servant.

NEIL DEGRASSE TYSON
Astrophysicist, American Museum of Natural History

••• CHAPTER TEN

THE NEXT FIFTY YEARS IN SPACE*

It would be hard to discuss the next fifty years in space without some reflection on the previous fifty. I happen to have been born the same week NASA was founded, in early October 1958. That means my earliest awareness of the world took place in the 1960s, during the Apollo era. It was also a turbulent decade internationally, and America was no exception. We were at war in Southeast Asia, the civil rights movement was under way, assassinations were taking place, and NASA was heading for the Moon.

At the time, it seemed clear that the astronauts, whatever criteria were used to select them, would never have included me. The astronauts were drawn from the military—all but two of them. One was Neil Armstrong, a civilian test pilot and aeronautical engineer—the commander of Apollo 11 and the first human to step foot on the Moon. The other was Harrison Schmitt, a geologist, the only scientist to go to the Moon. Schmitt was the lunar module pilot of Apollo 17, America's last Moon mission.

Perhaps the most turbulent year of that turbulent decade was 1968, yet that's the year Apollo 8 became the first craft ever to leave low Earth orbit and go to the Moon. That journey took place in December, at the end of an intense and bloody year. During Apollo 8's orbit, its astronauts took the most recognized photograph in the history of the

---

* Adapted from closing keynote speech at "50 Years of the Space Age," a celebration sponsored by the International Astronautical Federation, UNESCO HQ, Paris, March 21, 2007.

world. As the spacecraft emerged from behind the far side of the Moon, they pulled out the camera, looked through the window of the command module, and captured Earth rising over the lunar landscape. This widely published image, titled *Earthrise,* presented Earth as a cosmic object, aloft in the sky of another cosmic object. It was simultaneously thrilling and humbling, beautiful and also a little scary.

By the way, the title *Earthrise* is a bit misleading. Earth has tidally locked the Moon, which means that the Moon eternally shows only one side to us. The urge is strong to presume that Earth rises and sets for observers on the Moon just as the Moon rises and sets for observers on Earth. But as seen from the Moon's near side, Earth never rises. It's just always there, floating in the sky.

Everybody remembers the 1960s as the era of the right stuff, but it had its share of robotic missions as well. The first rovers on the Moon were Russian: Luna 9 and Luna 13. America's Ranger 7 was the first US spacecraft to photograph the Moon's surface. But those go unremembered by the public, even though they were our robotic forebearers in space, because there was a much bigger story being told: only when human emissaries were doing the exploring did people feel a vicarious attachment to the dramas unfolding on the space frontier.

Because I grew up in America, I took for granted that, by and large, everybody thinks about tomorrow, next year, five years from now, ten years from now. It's a popular pastime. If you say to someone, "So, what are you up to?" they're not going to tell you what they're doing today. No, they're going to tell you what they're planning: "I'm saving to go on a trip to the Caribbean," or "We're going to buy a bigger house," or "We're going to have two more kids." People are envisioning the future.

Americans are not alone in this, of course. But in some countries I visit, I speak to people who do not think about the future. And any country where people do not think about the future is a country without a space program. Space, I have learned, is a frontier that keeps you dreaming about what might get discovered tomorrow—a fundamental feature of being human.

Around the world and across time, every people and every culture—

even those with no written language—has some sort of story that accounts, mythologically or otherwise, for its existence and its relationship to the known universe. These are not new questions. These are old questions. This is an old quest.

Humans are one of very few animals that are perfectly happy sleeping on their backs. Also, we sleep at night. What happens if you wake up from sleeping at night on your back? You see the stars. It is possible that, of all the animals in the history of life on Earth, we may be uniquely curious about the sky, and so perhaps we should not be surprised that we wonder about our place in the cosmos.

Space Tweet #9

The night is our day. Marry astronomers—you'll always know where they are at night

Jul 14, 2010 6:08 AM

Today when we think of distant objects in space, we make plans to go there. We've gone to the Moon. We talk about the possibility of going to Mars. The twentieth century, of course, was the first in which the methods and tools of science—and particularly the methods and tools of space exploration—enabled us to answer age-old questions without reference to mythological sources: Where did we come from? Where are we going? Where do we fit in the universe? Many of our answers have come not simply because we went to the Moon or to some other celestial object but because space offers us places from which to access the rest of the cosmos.

Most of what the universe wants to tell us doesn't reach Earth's surface. We would know nothing of black holes were it not for telescopes launched into space. We would know nothing of various explosions in the universe that are rich in X-rays, gamma rays, or ultraviolet. Before we had vistas in space—telescopes, satellites, space probes—that enabled us to conduct astrophysical studies without interference from Earth's atmosphere, which we normally think of as transparent, we were almost blind to the universe.

When I think of tomorrow's space exploration, I don't think of low Earth orbit—altitudes less than about two thousand kilometers. In the 1960s that was a frontier. But now low Earth orbit is routine. It can still be dangerous, but it isn't a space frontier. Take me somewhere new. Do something more than drive around the block.

Yes, the Moon is a destination. Mars is a destination. But the Lagrangian points are destinations too. Those are where gravitational and centrifugal forces balance in a rotating system such as Earth and the Moon or Earth and the Sun. At destinations such as those, we can build things. We already have some experience, brought by building the International Space Station, which is bigger than most things ever conceived or constructed on Earth.

If you ask me, "What is culture?" I would say it is all the things we do as a nation or group or inhabitants of a city or region, yet no longer pay attention to. It's the things we take for granted. I'm a New Yorker, and so, for example, I no longer notice when I walk past a seventy-story building. Yet every tourist who comes to New York City from any place in the world is continually looking up. So I ask myself, What do people elsewhere take for granted in their own cultures?

Sometimes it's the simple things. Last time I visited Italy, I went to a supermarket and saw an entire aisle of pasta. I had never seen that before. There were pasta shapes that never make it to the United States. So I asked my Italian friends, "Do you notice this?" And they said no, it was simply the pasta aisle. In the Far East, there are entire aisles of rice, with choices undreamt of in America. So I asked a friend who wasn't born in America, "What's in our supermarkets that you think I no longer notice?" And she said, "You have an entire aisle of ready-to-eat breakfast cereals." To me, of course, that's just the cereal aisle. We have entire aisles full of soft drinks: Coke and Pepsi and all their derivatives. Yet that's just the soda aisle to me.

Where am I going with these examples? In America, everyday items incorporate icons from the space program. You can buy refrigerator magnets in the shape of the Hubble Space Telescope. You can buy boxes of bandages decorated not only with Spider-Man and Superman and Barbie but also with stars and moons and planets that glow in the dark. You can buy pineapple slices cut into Cosmic Fun Shapes. And for car names, the cosmos ranks second after geographic locations. This is the space component of culture that people no longer notice.

Space Tweet #10
Tasty Cosmos: Mars bar, Milky Way bar. MoonPie, Eclipse gum, Orbit gum, Sunkist, Celestial Seasonings. No food named Uranus
Jul 10, 2010 11:28 AM

Several years ago I served on a commission whose task was to analyze the future of the US aerospace industry—which had been falling on hard times, in part because of the success of Airbus in Europe and Embraer in Brazil. We went around the world to explore the economic climate in which American industries are functioning, so that we could advise Congress and the aerospace industry how to restore the leadership, or at least the competitiveness, that they (and we all) may once have taken for granted.

So we visited various countries in Western Europe and worked our way east. Our last stop was Moscow. One of the places we visited was Star City, a training center for cosmonauts where you'll find a striking monument in honor of Yuri Gagarin. Following the usual introductory platitudes and a morning shot of vodka, the director of Star City just sat back, loosened his tie, and spoke longingly of space. His eyes sparkled, as did mine, and I felt a connection I did not feel in England or France or Belgium or Italy or Spain.

That connection exists, of course, because our two nations, for a brief moment in history, directed major resources toward putting people into space. Having engaged in that endeavor has worked its way into both Russian and American culture, so that we don't conceive of life without it. My camaraderie with Star City's director made me think about what the world would be like if every country were engaged in that enterprise. I imagined our being connected with one another on a higher plane—beyond economic and military conflicts, beyond war altogether. I wondered how two nations with such deep, shared dreams about human presence in space could have remained such long-standing adversaries in the post–World War II era.

I have another example of space becoming part of culture. Three years ago, when NASA announced that an upcoming servicing mission to fix the Hubble Space Telescope might be canceled, it became big news in America. Do you know who played the biggest role in reversing that decision? Not the astrophysicists. It was the general public. Why? They had beautified their walls, computer screens, CD covers, guitars, and high-fashion gowns with Hubble images, and so in their own way they had become vicarious participants in cosmic discovery. The public took ownership of the Hubble Space Telescope, and eventually, after a slew of editorials, letters to the editor, talk-show discussions, and congressional debates, the funding was restored. I do not know of another time in the history of science when the public took ownership of a scientific instrument. But it happened then and there, marking the ascent of the Hubble into popular American culture.

I won't soon forget the deep feeling of commonality I had while sitting around schmoozing at Star City with members of the Russian space community. If the whole world shared such experiences, we would then have common dreams and everybody could begin thinking about tomorrow. And if everybody thinks about tomorrow, then someday we can all visit the sky together.

Space Tweet #11

Would a NASA reality show "Lunar Shore" be more popular than "Jersey Shore"? Civilization's future depends on that answer

May 16, 2011 8:18 AM

••• CHAPTER ELEVEN

SPACE OPTIONS

Podcast interview with Julia Galef and Massimo Pigliucci for *Rationally Speaking**

Julia Galef: Our guest in the studio today is Neil deGrasse Tyson, an astrophysicist and the director of the Hayden Planetarium. Neil is joining us to talk about the status of the space program today—what are its current goals, and what practical benefit does the space program have for our society? And to the extent that it doesn't have practical benefits, what are the justifications for spending taxpayer money on it—or on any other science without applied benefit?

Neil deGrasse Tyson: Let me remind some listeners, or alert them perhaps for the first time, what it is we're talking about. The Obama administration, in the new NASA budget, made some fundamental changes to the portfolio of NASA's ambitions. Some are good; some are neutral; some have been heavily criticized. The one that has had hardly any resistance, and was broadly praised, was the urge to get NASA out of low Earth orbit and to cede that activity to private enterprise.

Typically, the way our government has birthed new industries is to make the initial investments before capital markets can value them. That's where the high risk lies. Innovative ideas become inventions. Inventions become patents. Patents earn money. Only when risks are managed and understood do capital markets take notice. Right now, plenty of business goes on in low

---

* Adapted from "Neil deGrasse Tyson and the Need for a Space Program," interview with Massimo Pigliucci and Julia Galef of New York City Skeptics, for "Rationally Speaking: Exploring the Borderlands between Reason and Nonsense"; released March 28, 2010, at http://www.rationallyspeakingpodcast.org.

Earth orbit—all the consumer products that thrive on GPS, direct TV, other satellite communications. These are all commercial markets. So the thinking is to get NASA back on the frontier, where it belongs.

Massimo Pigliucci: Speaking of low Earth orbit, what exactly has the space station been doing up there?

NDT: Even more than research in Antarctica, the International Space Station is the prime example of international cooperation—the largest in human history, aside from the waging of world wars.

Multiple countries have gone down to Antarctica to do collaborative science research. And no one is making land grabs, maybe because no one wants to live there. So that helps in the collaboration: no one wants to be the King of Nothing. Antarctica is not only a beautiful place but also a unique location for conducting certain kinds of science—in part because it's cold, so there's low moisture in the air. And the South Pole happens to be at high elevation, so you're above layers of atmosphere that would otherwise interfere with your view of the night sky. As a result, astrophysics thrives at the South Pole.

The point is, just as Antarctica is an area of considerable international collaboration, so too is the International Space Station. It also demonstrates that we can build big things in space. We once thought that a telescope or some other piece of hardware required a surface on which to build it. But where there's a surface, there's gravity—which means the weight of the system requires structural support. But in orbit everything is weightless, permitting the building of huge structures that would be inherently unstable on Earth's surface.

MP: But would you therefore make an exception for the International Space Station, in terms of this issue of privatization as opposed to government funding of research?

NDT: You wouldn't necessarily privatize the space station itself right now, but you'd certainly privatize access to it. You'd sell the trips there. Why not? That's really where privatization would first reveal itself, according to the new plan. And no one's complaining about that. Where Obama got in a little bit of hot water was his cancellation of the NASA plan to return to the Moon.

The Moon is an interesting target. First, it's nearby. And having already been there means we can go there now with greater confidence of success, whereas a round trip to Mars involves dangers both known and unknown. Sending astronauts outside the protective blanket of Earth's magnetic field would leave them vulnerable to ionizing radiation from solar flares, which generate high-energy charged particles that can enter the body and ionize its atoms.

MP: So would you see a possible Moon station as a stepping-stone toward a Mars mission?

NDT: No, because if you're going to Mars, generally you don't want to go somewhere else first, because it takes energy to slow down, land, and take off again. Slowing down requires fuel. If the Moon had an atmosphere, you could use it to slow down, just as the space shuttle does as it returns to Earth. That's why it needs those famous tiles that dissipate the heat of reentry. If we didn't have a way to dump the energy of motion, the shuttle would be unable to stop.

Space Tweet #12
Just an FYI: If you blow-torch a shuttle tile to red-hot, in time it takes to put down the torch, tile is back to room temp
Mar 9, 2011 11:34 AM

Plus, do you bring all your resources with you? If you're taking a road trip to California, do you attach a supertanker to your car? Do you bring along a farm? No, you rely on the fact that there's a string of Quik Marts between here and California, so that you can refuel and buy food.

A long-term goal for living and working in space would be to exploit the resources that are already there. Obama's National Space Policy does say we should continue to do research on launch vehicles and rocket technologies that will one day get us to Mars, but when that day should come was not specified. And that's what makes space enthusiasts uncomfortable.

If we were choosing whether to go to the Moon or to Mars, most scientists—there are some key vocal exceptions, but I'm talking about most scientists, myself included—would pick Mars. It has plenty of evidence for a history of running water and enticing evidence for liquid water laying recent tracks within the soils. It also has methane, effusing its way out of a cliff face. What drives scientists to choose Mars is not just its fascinating geology (though perhaps we should call it marsology, since "geo-" means Earth). Deep down in our quest to know these planetary surfaces is our ongoing search for life, because every place on Earth where there's liquid water, there's life.

JG: Can you talk about the advantage of putting a human on Mars, as opposed to robotic exploration of Mars?

NDT: There's no advantage. That's the short answer. But let me provide some nuance. It costs anywhere from twenty to fifty times more money to send a human to a space destination than it does to send a robot. Say you're a geologist, and I tell you, "I could send you to Mars with your rock hammer and maybe a few machines to make measurements. I can do that once, *or* I can fund thirty different rovers that can be placed anywhere you choose on the Martian surface, and they'll carry the machines that I'd otherwise be giving to you." Which would you pick?

MP: It seems like a no-brainer to me.

NDT: Scientifically, it's a no-brainer. That's the point. It's because of the price difference that any scientist interested in scientific results would not, could not, with a clear conscience, send a human there. That leaves two options. Either you seriously lower the cost of sending humans there, so that it's competitive with sending robots, or you send a person regardless of the cost, because a person can do in a few minutes what it might take a rover all day to do. And that's because the human brain is more intuitive about what it's looking at than is the robot you've programmed. A program represents a subset of what you are, but it's still not you. And if you're the programmer, can you make a computer more intuitive than you are? I'll leave that one for you philosophers.

MP: Before the show we were talking about something very pertinent to this topic: how extremely large and expensive projects got funded historically.

NDT: There are really just three justifications for spending large portions of state wealth—three drivers. One of them is praise of royalty and deity:

ould potentially spark another influx and interest in fund-
ration.

tnik moment."

a good name for it. But the kind of research that might be justi-
t kind of reason might not be the best kind of scientific research.

ience alone has never been a driver of expensive projects. Below a level, depending on the wealth of a nation, money can be spent on e without heavy debate. For example, the price tag for the Hubble e Telescope, over all its years, is about $10 to $12 billion—less than $1 ion per year. That's comfortably below the radar of criticism for a science oject or for a project not based on the economy or war. Raise the cost of project above $20 billion to $30 billion, and if there's not a weapon at the other end of the experiment, or you won't see the face of God, or oil wells aren't to be found, it risks not getting funded. That's what happened with the Superconducting Super Collider. America was going to have the most powerful particle accelerator in the world; it was conceived in the late 1970s and funded in the mid-1980s. Then 1989 comes around. What happens? Peace breaks out.

MP: I hate when that happens!

JG: It's so inconvenient!

NDT: When you're at war, money flows like rivers. In 1945 physicists basically won the war in the Pacific with the Manhattan Project. Long before the bomb, and continuing through the entire Cold War, America sustained a fully funded particle physics program. Then the Berlin wall comes down in 1989, and within four years the entire budget for the Super Collider gets canceled.

What happens now? Europe says, "We'll take the mantle." They start building the Large Hadron Collider at CERN, the European Organization for Nuclear Research, and now we're standing on our shores and looking across the pond, crying out, "Can we join? Can we help?"

MP: I remember an interesting exchange from those hearings you're talking about. One senator who was evaluating the continued expense for the

activities undertaken in part out o
of the power for which you'r

MP: We could ask the Pope to

NDT: In principle, yes. However, w
commonly undertake such activities.
found. One is the promise of economic
I think of the pair as the I-don't-want-to-d
to-die-poor driver.

We all remember President Kennedy saying,
should commit itself to achieving the goal, before
landing a man on the moon and returning him safely
are powerful words; they galvanized the ambitions of a na
a speech given to a joint session of Congress on May 25, 1
weeks after the Soviet Union successfully launched Yuri Gagari
orbit—the first person to get there. Kennedy's speech was a reacti
fact that the United States did not yet have a "man-rated" rocket, m
a rocket safe enough for human spaceflight. To put a satellite in space,
might be willing to experiment with cheaper components or design tha
you'd use for putting a person up there.

A few paragraphs earlier in that same speech, Kennedy says, "If we are to win the battle that is now going on around the world between freedom and tyranny, the dramatic achievements in space which occurred in recent weeks should have made clear to us all, as did the Sputnik in 1957, the impact of this adventure on the minds of men everywhere, who are attempting to make a determination of which road they should take." This was a battle cry against communism.

MP: It was a political statement.

NDT: Period. He could have said, "Let's go to the Moon: what a marvelous place to explore!" But that's not enough to get Congress to write the check. At some point, somebody's got to write a check.

JG: Right. The Soviet Union was the catalyst then, and China is the catalyst now. China's space program is developing, right? And in the next ten or fifteen years, China may be poised to rival us as the superpower of the

Super Collider said to Steven Weinberg, a physicist testifying before Congress, "Unfortunately, one of the problems is that it's hard for me to justify this expense to my constituents, because, after all, nobody eats quarks." And then Weinberg, in his typical fashion, pretended to do a little calculation on the piece of paper in front of him and, as I remember it, said something along the lines of, "Actually, Senator, by my calculation, you just ate a billion billion billion quarks this morning for breakfast." In any case, the bottom line is that large basic-research projects get funded only if they piggyback on, as you said, the big three.

NDT: Either they have to piggyback on one of them or come in below the funding threshold for getting scrutinized.

MP: Somebody may reasonably ask, "Should it be otherwise?" In some sense, the senator brought up a good question: How do I justify this to my constituents?

NDT: I claim that even if Weinberg had said, "At the end of this, you'll get great technological spin-offs," it would still have been canceled. He would have had to say, "At the end of this, you'll have a weapon that protects the country." There's a famous reply, I don't remember who said it to whom, but it would have played well here. The senator says to the scientist, "What aspects of this project will help in the defense of America?"—there it is, plainly stated: the question of war—and the scientist replies, "Senator, I don't know how it can help in the defense of America, other than to ensure that America is a country worth defending."

MP: And that, as you know, is a great argument that doesn't fly.

NDT: Yes, it makes a good headline, but no, it doesn't garner the funding. Unless we're going to believe we're a fundamentally different kind of population and culture than those that have preceded us for the past five thousand years, I'm going to take my cue from the history of major funded projects and say that if we want to go to Mars, we'd better find either an economic driver or a military driver for it. Sometimes I half-joke about this and say, "Let's get China to leak a memo that says they want to build military bases on Mars. We'd be on Mars in twelve months."

JG: Do you think there's any case to be made for the fact that so many scientific discoveries that end up being incredibly useful and practical were discovered accidentally, in the course of exploratory research or completely unrelated research—that the discoverers got lucky? Can we make that case for space exploration?

NDT: That's an excellent question. But no, because the time delay between a serendipitous scientific discovery on the frontier and the fully developed product that has been engineered, designed, and marketed is typically longer than the reelection cycles of those who allocate money. Therefore it does not survive. You can't get politicians to decide to invest this way, because it's irrelevant to the needs of their constituencies. So I don't think we'll ever go to Mars unless we can find an economic or a military reason for doing so.

By the way, I know how to justify the $100 billion. But my pitch takes longer than what's called the "elevator conversation" with the member of Congress, where you get only thirty seconds to make your case, and it's your only chance—go! I need maybe three minutes.

JG: You could stop the elevator.

MP: Or, if you wanted to make the point to the general public rather than the congressman, you could say, "Here are good reasons to fund space exploration or basic scientific research in astrophysics. It's not just my curiosity or my wanting to be paid to do things I like."

NDT: In fact, we *are* funding basic research in astrophysics. But my conversation with you is about the manned space program. That's where the expense comes in. That's where all your budget options come in above the funding threshold for heavy scrutiny, and you have no choice but to appeal to these great drivers in the history of culture. As far as basic research goes, we've got the Hubble telescope; we're going to have a laboratory on Mars in a few years; we have the spacecraft Cassini in orbit around Saturn right now, observing the planet and its moons and its ring systems. We've got another spacecraft on its way to Pluto. We've got telescopes being designed and built that will observe more parts of the electromagnetic spectrum. Science is getting done. I wish there was more of it, but it's getting done.

MP: But not the Large Hadron Collider, which is getting done by the Europeans.

JG: There's one other potential case for space travel that we haven't really talked about. Earlier you alluded to the idea that if we become a spacefaring people, we might need to use the Moon and Mars as a sort of Quik Mart. Do you think we could make the practical case that we need to venture out into space because Earth will at some point become uninhabitable?

NDT: There are many who make that case. Stephen Hawking is among them; J. Richard Gott at Princeton is another. But if we acquire enough know-how to terraform Mars and ship a billion people there, surely that know-how will include the capacity to fix Earth's rivers, oceans, and atmosphere, as well as to deflect asteroids. So I don't think escaping to other planets is necessarily the most expedient solution to protecting life on Earth.

••• CHAPTER TWELVE

PATHS TO DISCOVERY*

From the Discovery of Places to the Discovery of Ideas

In how many ways does society today differ from that of last year, last century, or last millennium? The list of medical and scientific achievements would convince anybody that we live in special times. It's easy to notice what is different; the challenge is to see what has remained the same.

Behind all the technology, we're still human beings, no more or less so than participants in all the rest of recorded history. In particular, some of the basic forces in organized society change slowly, if at all; contemporary humans still exhibit basic behaviors. We climb mountains, wage war, vie for sex, seek entertainment, and long for economic and political power. Complaints about the demise of society and the "youth of today" also tend to be timeless. Consider this pronouncement, inscribed on an Assyrian tablet circa 2800 B.C.:

> Our earth is degenerate these days . . . bribery and corruption abound, children no longer obey their parents, every man wants to write a book, and the end of the world is evidently approaching.

The urge to climb a mountain may not be shared by everyone, but the urge to discover—which might drive some people to climb mountains and

* Adapted from "Paths to Discovery," chapter 19 in Richard W. Bulliet, ed., *The Columbia History of the 20th Century* (New York: Columbia University Press, 1998).

others to invent methods of cooking—does seem to be shared, and that tendency has been uniquely responsible for changes in society across the centuries. Discovery is the only enterprise that builds upon itself, persists from generation to generation, and expands human understanding of the universe. This is true whether the boundary of your known world is the other side of the ocean or the other side of the galaxy.

Discovery provokes comparisons between what you already know to exist and what you have just discovered. Successful prior discoveries often help dictate how subsequent discoveries unfold. To find something that has no analog to your own experience constitutes a personal discovery. To find something with no analog to the sum of the world's known objects, life-forms, practices, and physical processes constitutes a discovery for all of humanity.

The act of discovery can take many forms beyond "look what I've found!" Historically, discoverers were people who embarked on long ocean voyages to unknown places. When they reached a destination, they could see, hear, smell, feel, and taste up close what was inaccessible from far away. Such was the Age of Exploration through the sixteenth century. But once the world had been explored and the continents mapped, human discovery began to focus not on voyages but on concepts.

The dawn of the seventeenth century saw the near-simultaneous invention of what are arguably the two most important scientific instruments ever conceived: the microscope and the telescope. (Not that this should be a measure of importance, but among the eighty-eight constellations are star patterns named for each: Microscopium and Telescopium.) The Dutch optician Antoni van Leeuwenhoek subsequently introduced the microscope to the world of biology, while the Italian physicist and astronomer Galileo Galilei turned a telescope of his own design to the sky. Jointly, they heralded a new era of technology-aided discovery, whereby the capacities of the human senses could be extended, revealing the natural world in unprecedented, even heretical, ways. Bacteria and other simple organisms whose existence could be revealed only through a microscope yielded knowledge that transcended the prior limits of human experience. The fact that Galileo revealed the Sun to have spots, the planet Jupiter to have satellites, and Earth not to be the center of all celestial motion was enough to unsettle centuries of

Aristotelian teachings by the Catholic Church and to put Galileo under house arrest.

Telescopic and microscopic discoveries defied "common sense." They forever changed the nature of discovery and the paths taken to achieve it; no longer would common sense be accepted as an effective tool of intellectual investigation. Our unaided five senses were shown to be not only insufficient but untrustworthy. To understand the world required trustworthy measurements—which might not agree with one's preconceptions—derived from experiments conducted with care and precision. The scientific method of hypothesis, unbiased testing, and retesting would rise to significance and continue unabated thenceforth, unavoidably shutting out the ill-equipped layperson from modern research and discovery.

Incentives to Discovery

Travel was the method of choice for most historic explorers because technology had not yet progressed to permit discovery by other means. Apparently it was so important for European explorers to discover something that the places they found were declared "discovered"—and ceremonially planted with flags—even when indigenous peoples were there in great numbers to greet them on the shores.

What drives us to explore? In 1969, the Apollo 11 astronauts Neil Armstrong and Buzz Aldrin Jr. landed, walked, and frolicked on the Moon. It was the first time in history that humans had landed on the surface of another world. Being Westerners as well as discoverers, we immediately fell back to our old imperialist ways—the astronaut-emissaries planted a flag—but this time no natives showed up to greet us. And the flag needed to have a stick inserted along its upper edge to simulate the effects of a supportive, photo-friendly breeze on that barren, airless world.

The lunar missions are generally considered to be humanity's greatest technological achievement. But I would propose a couple of modifications to our first words and deeds on the Moon. Upon stepping onto the lunar surface, Neil Armstrong said, "That's one small step for [a] man, one giant leap for mankind" and then proceeded to plant the American flag in lunar soil. If indeed his giant leap was for "mankind," perhaps the flag should have been that of the United Nations. If he had been politically honest, he would have referred to "one giant leap for the United States of America."

The revenue stream that fed America's era of space-age discovery derived from taxpayers and was motivated by the prospect of military conflict with the Soviet Union. Major funded projects require major motivation. War is a preeminent motivator, and was largely responsible for projects such as the Great Wall of China, the atomic bomb, and the Soviet and American space programs. Indeed, as a result of two world wars within thirty years of each other and the protracted Cold War that followed, scientific and technological discovery in the twentieth century was accelerated in the West.

A close second in incentives for major funded projects is the prospect of high economic return. Among the most notable examples are the voyages of Columbus, whose funding level was a nontrivial fraction of Spain's gross national product, and the Panama Canal, which made possible in the twentieth century what Columbus had failed to find in the fifteenth—a shorter trading route to the Far East.

Space Tweet #13
Columbus took three months to cross the Atlantic in 1492. The Shuttle takes 15 minutes
May 16, 2011 9:30 AM

When major projects are driven primarily by the sheer quest to discover, they stand the greatest chance of achieving major breakthroughs—that's what they're designed to do—but the least chance of being adequately funded. The construction of a superconducting supercollider in the United States—an enormous (and enormously expensive) underground particle accelerator that was to extend human understanding of the fundamental forces of nature and the conditions in the early universe—never got past a big hole in the ground. Perhaps that shouldn't surprise us. With a price tag of more than $20 billion, its cost was far out of proportion to the expected economic returns from spin-off technologies, and there was no obvious military benefit.

When major funded projects are driven primarily by ego or self-promotion, rarely do the achievements extend beyond architecture per se, as in the Hearst Castle in California, the Taj Mahal in India, and the Palace of Versailles in France. Such lavish monuments to individuals, which have always been a luxury of either a successful or an exploitative society, make unsurpassed tourist attractions but do not reach the level of discovery.

Most individuals cannot afford to build pyramids; a mere handful of us get to be the first on the Moon or the first anywhere. Yet that doesn't seem to stop the desire to leave one's mark. Like animals that delineate territory with growls or urine, when flags are unavailable ordinary people leave a carved or painted name instead—no matter how sacred or revered the discovered spot may be. If the Apollo 11 had forgotten to take along the flag, the astronauts just might have chiseled into a nearby boulder "NEIL & BUZZ WERE HERE—7/20/69." In any case, the space program left behind plenty of evidence on each visit: all manner of hardware and other jetsam, from golf balls to automobiles, is scattered on the Moon's surface as testament to the six Apollo missions. The litter-strewn lunar soil simultaneously represents the proof and the consequences of discovery.

Amateur astronomers, who monitor the sky far more thoroughly than anybody else, are especially good at discovering comets. The prospect of getting something named after oneself is strong motivation: to discover a bright comet means the world will be forced to identify it with your name. Well-known examples include Comet Halley, which needs no introduction; Comet Ikeya-Seki, perhaps the most beautiful comet of the twentieth century, with its long and graceful tail; and Comet Shoemaker-Levy 9, which plunged into Jupiter's atmosphere in July 1994, within a few days of the twenty-fifth anniversary of the Apollo 11 Moon landing. Although among the most famous celestial bodies of our times, these comets endured neither the planting of flags nor the carving of initials.

If money is the most widely recognized reward for achievement, then the twentieth century was off to a good start. A roll call of the world's greatest and most influential scientific discoveries can be found among the recipients of the Nobel Prize, endowed in perpetuity by the Swedish chemist Alfred Bernhard Nobel, from wealth accrued through the manufacture of armaments and the invention of dynamite. The impressive size of the prize—currently approaching a million and a half dollars—serves as a carrot for many scientists working in the fields of physics, medicine, and chemistry. The awards began in 1901, five years after Nobel's death—which is fortunate because scientific discovery was just then attaining a rate commensurate with an annual reward. But if the volume of published research in, say, astrophysics can be used as a barometer, then as much has been discovered in the past fifteen years as in the entire previous history of the field. Perhaps there will come a day when the Nobel science prizes will be awarded monthly.

Discovery and the Extension of Human Senses

If technology extends our muscle and brain power, science extends the power of our senses beyond inborn limits. A primitive way we can do better is to move closer and get a better look; trees can't walk, but they don't have eyeballs either. Among humans, the eye is often regarded as an impressive organ. Its capacity to focus near and far, to adjust to a broad range of light levels, and to distinguish colors puts it at the top of most people's list of desirable features. Yet when we take note of the many bands of light that are invisible to us, we are forced to declare humans to be practically blind—even after walking closer to get a better look. How impressive is our hearing? Bats clearly fly circles around us, given their sensitivity to pitch that exceeds our own by an order of magnitude. And if the human sense of smell were as good as that of dogs, then Fred rather than Fido might be sniffing out the drugs and bombs.

The history of human discovery is a history of the boundless desire to extend the senses, and it is because of this desire that we have opened new windows to the universe. Beginning in the 1960s with the early Soviet and NASA missions to the Moon and the solar system's planets, computer-controlled space probes—which we can rightly call robots—became (and still are) the standard tool for space exploration. Robots in space have several clear advantages over astronauts: they are cheaper to launch; they can be designed to perform experiments of very high precision without interference from a cumbersome pressure suit; and since they are not alive in any traditional sense of the word, they cannot be killed in a space accident. Nevertheless, until computers can simulate human curiosity and human sparks of insight, and until computers can synthesize information and recognize a serendipitous discovery when it stares them in the face, robots will remain tools designed to discover what we already expect to find. Unfortunately, profound insights into nature lurk behind questions we have yet to ask.

The most significant improvement of our feeble senses is the extension of our sight into the invisible bands of what is collectively known as the electromagnetic spectrum. In the late nineteenth century the German physicist Heinrich Hertz performed experiments that helped unify conceptually what had previously been considered unrelated forms of radiation. Radio waves, infrared, visible light, and ultraviolet were all revealed to be cousins in a family of light whose members simply differed in energy. The full spectrum, including all parts discovered after Hertz's work, runs from the low-energy part, called radio waves, and extends, in order of increasing energy, to microwaves, infrared, visible (comprising the "rainbow seven": red, orange, yellow, green, blue, indigo, and violet), ultraviolet, X-rays, and gamma rays.

Superman, with his X-ray vision, has few advantages over modern scientists. Yes, he is somewhat stronger than your average astrophysicist, but astrophysicists can now "see" into every major part of the electromagnetic spectrum. Lacking this extended vision, we would be not only blind but ignorant, because many astrophysical phenomena reveal themselves only in certain "windows" within the spectrum.

Let's peek at a few discoveries made through each window to the universe, starting with radio waves, which require very different detectors from those found in the human retina.

In 1931 Karl Jansky, then employed by Bell Telephone Laboratories and armed with a radio antenna he himself built, became the first human to "see" radio signals emanating from somewhere other than Earth. He had, in fact, discovered the center of the Milky Way galaxy. Its radio signal was so intense that if the human eye were sensitive only to radio waves, then the galactic center would be one of the brightest sources in the sky.

With the help of some cleverly designed electronics, it's possible to transmit specially encoded radio waves that can then be transformed into sound via an ingenious apparatus known as a radio. So, by virtue of extending our sense of sight, we have also, in effect, managed to extend our sense of hearing. Any source of radio waves—indeed, practically any source of energy at all—can be channeled so as to vibrate the cone of a speaker, a simple fact that is occasionally misunderstood by journalists. When radio emissions from Saturn were discovered, for instance, it was simple enough for astronomers to hook up a radio receiver equipped with a speaker; the signal was then converted to audible sound waves, whereupon more than one journalist reported that "sounds" were coming from Saturn, and that life on Saturn was trying to tell us something.

With much more sensitive and sophisticated radio detectors than were available to Karl Jansky, astrophysicists now explore not just the Milky Way but the entire universe. As a testament to the human bias toward seeing-is-believing, early detections of radio sources in the universe were often considered untrustworthy until they were confirmed by observations with a conventional telescope. Fortunately, most classes of radio-emitting objects also emit some level of visible light, so blind faith was not always required. Eventually radio telescopes produced a rich parade of discoveries, including quasars (loosely assembled acronym of "quasi-stellar radio source"), which are among the most distant and energetic objects in the known universe.

Gas-rich galaxies emit radio waves from their abundant hydrogen atoms (more than 90 percent of all atoms in the cosmos are hydrogen). Large arrays of electronically connected radio telescopes can generate very high resolution images of a galaxy's gas content, revealing intricate features such as twists, blobs, holes, and filaments. In many ways, the task of mapping galaxies is no different from that facing fifteenth- and sixteenth-century cartographers, whose renditions of continents—distorted though they were—represented a noble human attempt to describe worlds beyond one's physical reach.

Microwaves have shorter wavelengths and more energy than radio waves. If the human eye were sensitive to microwaves, you could see the radar emitted by the speed gun of a highway patrol officer hiding in the bushes, and microwave-emitting telephone relay towers would be ablaze with light. The inside of your microwave oven, however, would look no different than it does now, because the mesh embedded in the door reflects microwaves back into the cavity to prevent their escape. Your eyeballs' vitreous humor is thus protected from getting cooked along with your food.

Microwave telescopes, which were not actively used to study the universe until the late 1960s, enable us to peer into cool, dense clouds of interstellar gas that ultimately collapse to form stars and planets. The heavy elements in these clouds readily assemble into complex molecules whose signature in the microwave part of the spectrum is unmistakable because of their match with identical molecules that exist on Earth. Some of those cosmic molecules, such as NH_3 (ammonia) and H_2O (water), are household standbys. Others, such as deadly CO (carbon monoxide) and HCN (hydrogen cyanide), are to be avoided at all costs. Some remind us of hospitals—H_2CO (formaldehyde) and C_2H_5OH (ethyl alcohol)—and some don't remind us of anything: N_2H+ (dinitrogen monohydride ion) and HC_4CN (cyanodiacetylene). More than 150 molecules have been detected, including glycine, an amino acid that is a building block for protein and thus for life as we know it. We are indeed made of stardust. Antoni van Leeuwenhoek would be proud.

Without a doubt, the most important single discovery in astrophysics was made with a microwave telescope: the heat left over from the origin of the universe. In 1964 this remnant heat was measured in a Nobel Prize–winning observation conducted at Bell Telephone Laboratories by the physicists Arno Penzias and Robert Wilson. The signal from this heat is an omnipresent, omnidirectional ocean of light—often called the cosmic microwave background—that today registers about 2.7 degrees on the "absolute" temperature scale and is dominated by microwaves (though it radiates at all wavelengths). This discovery was serendipity at its finest. Penzias and Wilson had humbly set out to find terrestrial sources of interference with microwave communications; what they found was compelling evidence for the Big Bang theory. It's a little like fishing for a minnow and catching a blue whale.

Moving further along the electromagnetic spectrum, we get to infrared light. Invisible to humans, it is most familiar to fast-food fanatics, whose French fries are kept lukewarm under infrared lamps for hours before being purchased. Infrared lamps also emit visible light, but their active ingredient is an abundance of invisible infrared photons, which are readily absorbed by food. If the human retina were sensitive to infrared, then a midnight glance at an ordinary household scene, with all the lights turned off, would reveal all the objects that sustain a temperature in excess of room temperature: the metal that surrounds the pilot lights of a gas stove, the hot water pipes, the iron that somebody had forgotten to turn off after pressing crumpled shirt collars, and the exposed skin of any humans passing by. Clearly that picture is not more enlightening than what you would see with visible light, but it's easy to imagine one or two creative uses of such amplified vision, such as examining your home in winter to spot heat leaks from the window panes or roof.

As a child, I was aware that, at night, infrared vision would reveal monsters hiding in the bedroom closet only if they were warm-blooded. But everybody knows that your average bedroom monster is reptilian and cold-blooded. Thus, infrared vision would completely miss a bedroom monster, because it would simply blend in with the walls and door.

In the universe, the infrared window is particularly useful for probing dense clouds that contain stellar nurseries, within which infant stars are often enshrouded by leftover gas and dust. These clouds absorb most of the visible light from their embedded stars and re-radiate it in the infrared, rendering our visible-light window quite useless. This makes infrared especially useful for studying the plane of the Milky Way, because that's where the obscuration of visible light from our galaxy's stars is at its greatest. Back home, infrared satellite photographs of Earth's surface reveal, among other things, the paths of warm oceanic waters, such as the North Atlantic Drift current, which swirls west of the British Isles and keeps them from becoming a major ski resort.

The visible part of the spectrum is what humans know best. The energy emitted by the Sun, whose surface temperature is about six thousand degrees above absolute zero, peaks in the visible part of the spectrum, as does the sensitivity of the human retina, which is why our sight is so useful in the daytime. Were it not for this match, we could rightly complain that some of our retinal sensitivity was being wasted.

We don't normally think of visible light as penetrating, but light passes mostly unhindered through glass and air. Ultraviolet, however, is summarily absorbed by ordinary glass. So, if our eyes were sensitive only to ultraviolet, windows made of glass would not be much different from windows made of brick. Stars that are a mere four times hotter than the Sun are prodigious producers of ultraviolet light. Fortunately, such stars are also bright in the visible part of the spectrum, which means that their discovery has not depended on access to ultraviolet telescopes. Since our atmosphere's ozone layer absorbs most of the ultraviolet and X-rays that impinge upon it, a detailed analysis of very hot stars can best be obtained from Earth orbit or beyond, which has become possible only since the 1960s.

As if to herald a new century of extended vision, the first Nobel Prize ever awarded in physics went to the German physicist Wilhelm Röntgen in 1901 for his discovery of X-rays. Cosmically, both X-rays and ultraviolet can indicate the presence of black holes—among the most exotic objects in the universe. Black holes are voracious maws that emit no light—their gravity is too strong for even light to escape—but their existence can be tracked by the energy emitted from heated, swirling gas nearby. Ultraviolet and X-rays are the predominant form of energy released by material just before it descends into the black hole.

It's worth remembering that the act of discovery does not require that you understand, either in advance or after the fact, what you've discovered. That's what happened with the cosmic microwave background. It also happened with gamma-ray bursts. Mysterious, seemingly random explosions of high-energy gamma rays scattered across the sky were first detected in the 1960s by satellites searching out radiation from clandestine Soviet nuclear-weapons tests. Only decades later did spaceborne telescopes, in concert with ground-based follow-up observations, show them to be the signature of distant stellar catastrophes.

Discovery through detection can cover a lot of territory, including subatomic particles. But one in particular virtually defies detection: the elusive neutrino. Whenever a neutron decays into an ordinary proton and an electron, a member of the neutrino clan springs into existence. Within the core of the Sun, for instance, two hundred trillion trillion trillion neutrinos are produced every second, and then pass directly out of the Sun as if it were not there at all. Neutrinos are extraordinarily difficult to capture because they have exceedingly minuscule mass and hardly ever interact with matter. Building an efficient, effective neutrino telescope thus remains an extraordinary challenge.

The detection of gravitational waves, another elusive window on the universe, would reveal catastrophic cosmic events. But as of this writing, these waves, predicted in Einstein's 1916 theory of general relativity as "ripples" in space and time, have not yet been directly detected from any source. A good gravitational-wave telescope would be able to detect black holes orbiting one another, and distant galaxies merging. One can even imagine a time in the future when gravitational events in the universe—collisions, explosions, collapsed stars—are routinely observed. In principle, we might one day see beyond the opaque wall of cosmic microwave background radiation to the Big Bang itself. Like Magellan's crew, who first circumnavigated Earth and saw the limits of the globe, we would then have reached and discovered the limits of the known universe.

Discovery and Society

As a surfboard rides a wave, the Industrial Revolution rode the eighteenth and nineteenth centuries on the crest of decade-by-decade advances in people's understanding of energy as a physical concept and a transmutable entity. Engineering technology replaces muscle energy with machine energy. Steam engines convert heat into mechanical energy; dams convert the gravitational potential energy of water into electricity; dynamite converts chemical energy into explosive shock waves. In a remarkable parallel to the way these discoveries transformed earlier societies, the twentieth century saw information technology ride the crest of advances in electronics and miniaturization, birthing an era in which computer power replaced mind power. Exploration and discovery now occurred on wafers of silicon, with computers completing in minutes, and eventually in moments, what would once have required lifetimes spent in calculations. Even so, we may still be groping in the dark, because as our area of knowledge grows, so does the perimeter of our ignorance.

What is the cumulative influence of all this technology and cosmic discovery on society, aside from creating more effective instruments of destruction and further excuses to wage war? The nineteenth and early twentieth centuries saw the development of transportation that did not rely on energy from domestic animals—including the bicycle, the railroad, the automobile, and the airplane. The twentieth century also saw the introduction of liquid fuel rockets (thanks in part to Robert Goddard) and spaceships (thanks in part to Wernher von Braun). The discovery of improved means of transportation was especially crucial to geographically large but habitable nations such as the United States. So important is transportation to Americans that the disruption of traffic by any means, even if it occurs in another country, can make headlines. On August 7, 1945, for example, the day after America killed some seventy thousand Japanese in the city of Hiroshima, with tens of thousands more deaths following soon afterward, the front page of the *New York Times* announced, "FIRST ATOMIC BOMB DROPPED ON JAPAN." A smaller headline, also on the front page, read, "TRAINS CANCELED IN STRICKEN AREA; Traffic Around Hiroshima Is Disrupted." I don't know for sure, but I would bet that day's Japanese newspapers did not consider traffic jams to be a top news item.

Technological change affected not only destruction, of course, but also domesticity. With electricity available in every domicile, it became worthwhile to invent appliances and machines that would consume this new source of energy. Among anthropologists, one of the broad measures of the advancement of society is its per-capita consumption of energy. Old traditions die hard, though. Lightbulbs were a substitute for candles, but we still light candles at special dinners; we even buy electric chandeliers studded with lightbulbs in the shape of candle flames. And of course car engines are measured in "horse" power.

The dependence on electricity, especially among urban Americans, has reached irreversible levels. Consider New York City during the blackouts of November 1965, July 1977, and August 2003, when this decidedly twentieth-century luxury temporarily became unavailable. In 1965, many people thought the world was going to end, and in 1977 there was widespread looting. (Each blackout allegedly produced "blackout babies," conceived in the absence of television and other technological distractions.) Apparently, our discoveries and inventions have gone from making life easier to becoming a requirement for survival.

Throughout history, discovery held risks and dangers for the discoverers themselves. Neither Magellan nor most of his crew remained alive to complete the round-the-world voyage in 1522. Most died of disease and starvation, and Magellan himself was killed by indigenous Filipinos who were not impressed with his attempts to Christianize them. Modern-day risks can be no less devastating. At the end of the nineteenth century, investigating high-energy radiation, Wilhelm Röntgen explored the properties of X-rays and Marie Curie explored the properties of radium. Both died of cancer. The three crew members of Apollo 1 burned to death on the launchpad in 1967. The space shuttle Challenger exploded shortly after launch in 1986, while space shuttle Columbia broke up on reentry in 2003, in both cases killing all seven crew members.

Sometimes the risks extend far beyond the discoverers. In 1905 Albert Einstein introduced the equation $E = mc^2$, the unprecedented recipe that interchanged matter with energy and ultimately begat the atomic bomb. Coincidentally, just two years before the first appearance of Einstein's famous equation, Orville Wright made the first successful flight in an airplane, the vehicle that would one day deliver the first atomic bombs in warfare. Shortly after the invention of the airplane, there appeared in one of the widely distributed magazines of the day a letter to the editor expressing concern over possible misuse of the new flying machine, noting that if an evil person took command of a plane, he might fly it over villages filled with innocent, defenseless people and toss canisters of nitroglycerin on them.

Wilbur and Orville Wright are, of course, no more to blame for the deaths resulting from military application of the airplane than Albert Einstein is to blame for deaths resulting from atomic bombs. For better or for worse, discoveries take their place in the public domain and are thus subject to patterns of human behavior that seem deeply embedded and quite ancient.

Discovery and the Human Ego

The history of human ideas about our place in the universe has been a long series of letdowns for everybody who likes to believe we're special. Unfortunately, first impressions have consistently fooled us—the daily motions of the Sun, Moon, and stars all conspire to make it look as though we are the center of everything. But over the centuries we have learned this is not so. There is no center of Earth's surface, so no culture can claim to be geometrically in the middle of things. Earth is not the center of the solar system; it is just one of multiple planets in orbit around the Sun, a revelation first proposed by Aristarchus in the third century B.C., argued by Nicolaus Copernicus in the sixteenth century, and consolidated by Galileo in the seventeenth. The Sun is about 25,000 light-years from the center of the Milky Way galaxy, and it revolves anonymously around the galactic center along with hundreds of billions of other stars. And the Milky Way is just one of a hundred billion galaxies in a universe that actually has no center at all. Finally, of course, owing to Charles Darwin's *Origin of Species* and *Descent of Man*, it is no longer necessary to invoke a creative act of divinity to explain human origins.

Scientific discovery is rarely the consequence of an instantaneous act of brilliance, and the revelation that our galaxy is neither special nor unique was no exception. The turning point in human understanding of our place in the cosmos occurred not centuries ago but in the spring of 1920, during a now-famous debate on the extent of the known universe, held at a meeting of the National Academy of Sciences in Washington, DC, at which fundamental questions were addressed: Was the Milky Way galaxy—with all its stars, star clusters, gas clouds, and fuzzy spiral things—all there was to the universe? Or were those fuzzy spiral things galaxies unto themselves, just like the Milky Way, dotting the unimaginable vastness of space like "island universes"?

Scientific discovery, unlike political conflict or public policy, does not normally emerge from party-line politics, democratic vote, or public debate. In this case, however, two leading scientists of the day, each armed with some good data, some bad data, and some sharpened arguments, went head to head at the Smithsonian's National Museum of Natural History. Harlow Shapley argued that the Milky Way constitutes the full extent of the universe, while Heber D. Curtis defended the opposing view.

Earlier in the century, both scientists had participated in a wave of discoveries derived primarily from classification schemes for cosmic objects and phenomena. With the help of a spectrograph (which breaks up starlight into its component colors the way raindrops break up sunlight into a rainbow), astrophysicists were able to classify objects not simply by their shape or outward appearance but by the detailed features revealed in their spectra. Even in the absence of full understanding of the cause or origin of a phenomenon, a well-designed classification scheme makes substantive deductions possible.

The nighttime sky displays a grab bag of objects whose classifications were not subject to much disagreement in 1920. Three kinds were especially relevant to the debate: the stars that are quite concentrated along the narrow band of light called the Milky Way, correctly interpreted by 1920 as the flattened plane of our own galaxy; the hundred or so titanic, roughly spherical globular star clusters that appear more frequently in just one direction of the sky; and third (or perhaps third and fourth), the inventory of fuzzy nebulae near the plane and spiral nebulae nowhere near the plane. Whatever else Shapley and Curtis intended to argue, they knew that those basic observed features of the sky could not be reasoned away. And although the data were scant, if Curtis could show that the spiral nebulae were distant island universes, then humanity would be handed the next chapter in its long series of ego-busting discoveries.

In a casual look at the night sky, stars appear uniformly spread in all directions along the Milky Way. But in fact, the Milky Way contains a mixture of stars and obscuring dust clouds that compromise lines of sight so that it becomes impossible to see the entire galaxy from within. In other words, you can't identify where you are in the Milky Way because the Milky Way is in the way. Nothing unusual there: the moment you enter a dense forest, you have no idea where you are within it (unless you carved your initials into a tree during a previous visit). The full extent of the forest is impossible to determine because the trees are in the way.

Astronomers of the day were fairly clueless as to how far away things are, and Shapley's estimates of distance tended to be quite generous, indeed excessive. Through various calculations and assumptions, he ended up with a galactic system more than 300,000 light-years in extent—by far the largest estimate ever made before (or since) for the size of the Milky Way. Curtis was unable to fault Shapley's reasoning but remained skeptical nonetheless, calling the assumption "rather drastic." Though based on the work of two leading theorists of the day, it was indeed rather drastic—and those theorists' relevant ideas would soon be discredited, leaving Shapley with overestimates in stellar luminosities and, as a result, overestimates in the distances to his favorite objects, the globular clusters.

Curtis remained convinced that the Milky Way galaxy was much smaller than suggested by Shapley, proposing that in the absence of definitive evidence to the contrary, "the postulated diameter of 300,000 light-years must quite certainly be divided by five, and perhaps by ten."

Who was right?

Along most paths from scientific ignorance to scientific discovery, the correct answer lies somewhere between the extreme estimates collected along the way. Such was the case here, too. Today, the generally accepted extent of the Milky Way galaxy is about 100,000 light-years—about three times Curtis's 30,000 light-years, and one-third Shapley's 300,000 light-years.

But that wasn't the end of it. The two debaters had now to reconcile the extent of the Milky Way with the existence of high-velocity spiral nebulae, whose distances were even more highly uncertain, and which seemed to avoid the galactic plane altogether, earning the Milky Way the spooky alternative name "Zone of Avoidance."

Shapley suggested that the spiral nebulae had somehow been created within the Milky Way and then forcibly ejected from their birthplace. Curtis was convinced that the spiral nebulae belonged to the same class of objects as the Milky Way itself, and proposed that a ring of "occulting matter" surrounded our galaxy—as is true of so many other spiral galaxies—and might be obliterating distant spirals from view.

At that point, if I were the moderator, I might have ended the debate, declared Curtis the winner, and sent everybody home. But there was further evidence at hand: the "novae," tremendously bright stars that occasionally, and very briefly, appear out of nowhere. Curtis contended that the novae formed a homogeneous class of objects that suggested "distances ranging from perhaps 500,000 light-years in the case of the Nebula in Andromeda, to 10,000,000 or more light-years for the more remote spirals." Given those distances, those island universes would be "of the same order of size as our own galaxy." Bravo.

Even though Shapley discounted the concept of the spiral nebulae as island universes, he no doubt wanted to appear open-minded. In his summary, which reads like a disclaimer, he entertained the possibility of other worlds:

> But even if spirals fail as galactic systems, there may be elsewhere in space stellar systems equal to or greater than ours—as yet unrecognized and possibly quite beyond the power of existing optical devices and present measuring scales. The modern telescope, however, with such accessories as high-power spectroscopes and photographic intensifiers, is destined to extend the inquiries relative to the size of the universe much deeper into space.

How right he was. Meanwhile, Curtis openly conceded that Shapley might be on to something with his hypothesis concerning the ejection of spiral nebulae, and in the course of that concession, Curtis unwittingly managed to reveal that we live in an expanding universe: "The repulsion theory, it is true, is given some support by the fact that most of the spirals observed to date are receding from us."

By 1925, a mere half decade later, Edwin Hubble had discovered that nearly all galaxies recede from the Milky Way at speeds in direct proportion to their distances. But it was self-evident that our galaxy, the Milky Way, was in the center of the expansion of the universe. Having been an attorney before becoming an astronomer, Hubble probably would have won any debate he might have had with other scientists, no matter what he argued, but he clearly could muster the evidence for an expanding universe with us at the center. In the context of Albert Einstein's general theory of relativity, however, the appearance of being at the center of an expanding fabric of space and time was a natural consequence of a four-dimensional cosmos, with time as number four. Given that description of the universe, the inhabitants of every galaxy would observe all other galaxies to be receding, not through space but as part of it, leading inescapably to the conclusion that Earthlings are neither alone nor special.

And the onward momentum toward insignificance continued with a vengeance.

In the 1920s and 1930s, physicists demonstrated that the fuel source in the Sun was the thermonuclear fusion of hydrogen into helium. In the 1940s and 1950s astrophysicists deduced the cosmic abundance of elements by describing in detail the sequence of thermonuclear fusion that unfolds in the cores of high-mass stars that explode at the end of their lives, enriching the universe with elements from all over the famed periodic table, the top five being hydrogen, helium, oxygen, carbon, and nitrogen. That very same sequence (except for helium, which is chemically inert) pops up when we look at the chemical constituents of human life. So, not only is our existence as human beings not special; neither are the ingredients of life itself.

So there you have it: the capsule summary of how cosmic discovery began by glorifying God, descended into glorifying human life, and ended up by insulting our collective human ego.

The Future of Discovery

When (or if) space ever becomes our final frontier, it will represent uncharted territories akin to those the ancient explorers dreamed of conquering. The coming voyages to space may be economically driven, for example by the intent to mine million-ton asteroids for their mineral resources. Or perhaps the voyages will be motivated by survival, spurred by the intent to spread the human species around the galaxy as much as possible so as to avoid total human extinction from a catastrophic, once-in-a-hundred-million-year collision with a comet or asteroid.

The golden era of space exploration was no doubt the 1960s. At that time, though, the significance of the space program was somewhat muddled in many urban centers because of widespread poverty, crime, and problem-ridden schools. Five decades later, the significance of the space program remains muddled in many urban centers because of widespread poverty, crime, and problem-ridden schools. But there's a fundamental difference. In the 1960s, discoveries in space were something that people looked forward to. Today many people—including me—are looking back at them.

I remember the day, and the moment, when the Apollo 11 astronauts stepped foot on the Moon. That landing, on July 20, 1969, was of course one of the twentieth century's greatest moments. Yet I found myself somewhat indifferent to the event—not because I couldn't appreciate its rightful place in human history, but because I had every reason to believe that trips to the Moon would soon take place monthly. Frequent Moon voyages were simply the next step; little did I know there would be a flurry of them in the twentieth century, followed by nothing for decades.

Yes, the funding stream for the space program had been primarily defense-driven. Cosmic dreams, and the innate human desire to explore the unknown, were of lesser import. But the word "defense" can be reinterpreted to mean something far more important than armies and arsenals. It can mean the defense of the human species itself. In July 1994 the equivalent of more than 200,000 megatons of TNT was deposited in Jupiter's upper atmosphere as comet Shoemaker-Levy 9 slammed into the planet. If that kind of collision happens on Earth while humanity is present, it would very likely result in the abrupt extinction of our species.

Defense of our existence mandates a very real agenda. To achieve it, we must acquire maximal understanding of Earth's climate and ecosystem, so as to minimize the risk of self-destruction, and we must colonize space in as many places as possible, thereby proportionally reducing the chance of species annihilation owing to a collision between Earth and an asteroid or comet discovered by an amateur astronomer.

The fossil record teems with extinct species. Many of them, before disappearing, thrived far longer than the current Earth tenure of *Homo sapiens*. Dinosaurs are extinct today because they did not build spacecraft. Were no funds available? Did their politicians lack foresight? More likely it was because their brains were tiny. And the absence of an opposable thumb didn't help either.

For humans to become extinct would be the greatest tragedy in the history of life in the universe—because the reason for it would be not that we lacked the intelligence to build interplanetary spacecraft, or that we lacked an active program of space travel, but that the human species itself turned its back and chose not to fund such a survival plan. Make no mistake: the path to discovery inherent in space exploration has become not a choice but a necessity, and the consequences of that choice affect the survival of absolutely everyone, including those who remain thoroughly unenlightened by the multitude of discoveries made by their own species throughout its time on Earth.

PART II

HOW

••• CHAPTER THIRTEEN

TO FLY*

> In ancient days two aviators procured to themselves wings. Daedalus flew safely through the middle air, and was duly honoured in his landing. Icarus soared upwards to the sun till the wax melted which bound his wings, and his flight ended in a fiasco. In weighing their achievements perhaps there is something to be said for Icarus. The classic authorities tell us, of course, that he was only "doing a stunt"; but I prefer to think of him as the man who certainly brought to light a serious constructional defect in the flying-machines of his day [and] we may at least hope to learn from his journey some hints to build a better machine.
>
> —Sir Arthur Eddington, *Stars & Atoms* (1927)

For millennia, the idea of being able to fly occupied human dreams and fantasies. Waddling around on Earth's surface as majestic birds flew overhead, perhaps we developed a form of wing envy. One might even call it wing worship.

You needn't look far for evidence. For most of the history of broadcast television in America, when a station signed off for the night, it didn't show somebody walking erect and bidding farewell; instead it would play the "Star Spangled Banner" and show things that fly, such as birds soaring or Air Force jets whooshing by. The United States even adopted a flying predator as a symbol of its strength: the bald eagle, which appears on the

* Adapted from "To Fly," *Natural History*, April 1998.

back of the dollar bill, the quarter, the Kennedy half dollar, the Eisenhower dollar, and the Susan B. Anthony dollar. There's also one on the floor of the Oval Office in the White House. Our most famous superhero, Superman, can fly upon donning blue pantyhose and a red cape. When you die, if you qualify, you might just become an angel—and everybody knows that angels (at least the ones who have earned their wings) can fly. Then there's the winged horse Pegasus; the wing-footed Mercury; the aerodynamically unlikely Cupid; and Peter Pan and his fairy sidekick, Tinkerbell.

Our inability to fly often goes unmentioned in textbook comparisons of human features with those of other species in the animal kingdom. Yet we are quick to use the word "hapless" as a synonym for "flightless" when describing a bird such as the dodo, which tends to find itself on the wrong end of evolutionary jokes. We did, however, ultimately learn to fly because of the technological ingenuity afforded by our human brains. And of course, while birds can fly, they are nonetheless stuck with bird brains. But this self-aggrandizing line of reasoning is somewhat flawed, because it ignores all the millennia that we were technologically flightless.

I remember as a student in junior high school reading that the famed physicist Lord Kelvin, at the turn of the twentieth century, had argued the impossibility of self-propelled flight by any device that was heavier than air. Clearly this was a myopic prediction. But one needn't have waited for the invention of the first airplanes to refute the essay's premise. One merely needed to look at birds, which have no trouble flying and, last I checked, are all heavier than air.

Space Tweet #14
USAirForce has styled bird wings as symbol. But we now fly at speeds that'd vaporize a bird, & in space, wings are useless
Sept 30, 2010 1:01 PM

If something is not forbidden by the laws of physics, then it is, in principle, possible, regardless of the limits of one's technological foresight. The speed of sound in air ranges from seven hundred to eight hundred miles per hour, depending on the atmospheric temperature. No law of physics prevents objects from going faster than Mach 1, the speed of sound. But before the sound "barrier" was broken in 1947 by Charles E. "Chuck" Yeager, piloting the Bell X-1 (a US Army rocket plane), much claptrap was written about the impossibility of objects moving faster than the speed of sound. Meanwhile, bullets fired by high-powered rifles had been breaking the sound barrier for more than a century. And the crack of a whip or the sound of a wet towel snapping at somebody's buttocks in the locker room is a mini sonic boom, created by the end of the whip or the tip of the towel moving through the air faster than the speed of sound. Any limits to breaking the sound barrier were purely psychological and technological.

During its lifetime, the fastest winged aircraft by far was the space shuttle, which, with the aid of detachable rockets and fuel tanks, exceeded Mach 20 on its way to orbit. Propulsionless on return, it fell back out of orbit, gliding safely down to Earth. Although other craft routinely travel many times faster than the speed of sound, none can travel faster than the speed of light. I speak not from a naiveté about technology's future but from a platform built upon the laws of physics, which apply on Earth as they do in the heavens. Credit the Apollo astronauts who went to the Moon with being the first to reach Earth's escape velocity—seven miles per second, the highest speed at which humans have ever flown, before or since. This is a paltry 1/250 of one percent of the speed of light. Actually, the real problem is not the moat that separates these two speeds but the laws of physics that prevent any object from ever achieving the speed of light, no matter how inventive your technology. The sound barrier and the light barrier are not equivalent limits on invention.

The Wright brothers of Ohio are, of course, generally credited with being "first in flight" at Kitty Hawk, North Carolina, as that state's license-plate slogan reminds us. But this claim needs to be further delineated. Wilbur and Orville Wright were the first to fly a heavier-than-air, engine-powered vehicle that carried a human being—Orville, in this case—and that did not land at a lower elevation than its takeoff point. Previously, people had flown in balloon gondolas and in gliders and had executed controlled descents from the sides of cliffs, but none of those efforts would have made a bird jealous. Nor would Wilbur and Orville's first trip have turned any bird heads. The first of their four flights—at 10:35 A.M. eastern time on December 17, 1903—lasted twelve seconds, at an average speed of 6.8 miles per hour against a 30-mile-per-hour wind. The Wright Flyer, as it was called, had traveled 120 feet, not even the length of one wing on a Boeing 747.

Even after the Wright brothers went public with their achievement, the media took only intermittent notice of it and other aviation firsts. As late as 1933—six years after Lindbergh's historic solo flight across the Atlantic—H. Gordon Garbedian ignored airplanes in the otherwise prescient introduction to his book *Major Mysteries of Science:*

> Present day life is dominated by science as never before. You pick up a telephone and within a few minutes you are talking with a friend in Paris. You can travel under sea in a submarine, or circumnavigate the globe by air in a Zeppelin. The radio carries your voice to all parts of the earth with the speed of light. Soon, television will enable you to see the world's greatest spectacles as you sit in the comfort of your living room.

But some journalists did pay attention to the way flight might change civilization. After the Frenchman Louis Blériot crossed the English Channel from Calais to Dover on July 25, 1909, an article on page three of the *New York Times* was headlined "FRENCHMAN PROVES AEROPLANE NO TOY." The article went on to delineate England's reaction to the event:

> Editorials in the London newspapers buzzed about the new world where Great Britain's insular strength is no longer unchallenged; that the aeroplane is not a toy but a possible instrument of warfare, which must be taken into account by soldiers and statesmen, and that it was the one thing needed to wake up the English people to the importance of the science of aviation.

The guy was right. Thirty-five years later, not only had airplanes been used as fighters and bombers in warfare but the Germans had taken the concept a notch further and invented the V-2 to attack London. Their vehicle was significant in many ways. First, it was not an airplane; it was an unprecedentedly large missile. Second, because the V-2 could be launched several hundred miles from its target, it basically birthed the modern rocket. And third, for its entire airborne journey after launch, the V-2 moved under the influence of gravity alone; in other words, it was a suborbital ballistic missile, the fastest way to deliver a bomb from one location on Earth to another. Subsequently, Cold War "advances" in the design of missiles enabled military power to target cities on opposite sides of the world. Maximum flight time? About forty-five minutes—not nearly enough time to evacuate a targeted city.

While we can say they're suborbital, do we have the right to declare missiles to be flying? Are falling objects in flight? Is Earth "flying" in orbit around the Sun? In keeping with the rules applied to the Wright brothers, a person must be onboard the craft and it must move under its own power. But there's no rule that says we cannot change the rules.

Knowing that the V-2 brought orbital technology within reach, some people got impatient. Among them were the editors of the popular, family-oriented magazine *Collier's*, which sent two journalists to join the engineers, scientists, and visionaries gathered at New York City's Hayden Planetarium on Columbus Day, 1951, for its seminal Space Travel Symposium. In the March 22, 1952, issue of *Collier's*, in a piece titled "What Are We Waiting For?" the magazine endorsed the need for and value of a space station that would serve as a watchful eye over a divided world:

> In the hands of the West a space station, permanently established beyond the atmosphere, would be the greatest hope for peace the world has ever known. No nation could undertake preparations for war without the certain knowledge that it was being observed by the ever-watching eyes aboard the "sentinel in space." It would be the end of the Iron Curtains wherever they might be.

We Americans didn't build a space station; instead we went to the Moon. With this effort, our wing worship continued. Never mind that Apollo astronauts landed on the airless Moon, where wings are completely useless, in a lunar module named after a bird. A mere sixty-five years, seven months, three days, five hours, and forty-three minutes after Orville left the ground, Neil Armstrong gave his first statement from the Moon's surface: "Houston, Tranquillity Base here. The Eagle has landed."

The human record for "altitude" does not go to anybody for having walked on the Moon. It goes to the astronauts of the ill-fated Apollo 13. Knowing they could not land on the Moon after the explosion in their oxygen tank, and knowing they did not have enough fuel to stop, slow down, and head back, they executed a single figure-eight ballistic trajectory around the Moon, swinging them back toward Earth. The Moon just happened to be near apogee, the farthest point from Earth in its elliptical orbit. No other Apollo mission (before or since) went to the Moon during apogee, which granted the Apollo 13 astronauts the human altitude record. (After calculating that they must have reached about 245,000 miles "above" Earth's surface, including the orbital distance from the Moon's surface, I asked Apollo 13 commander Jim Lovell, "Who was on the far side of the command module as it rounded the Moon? That single person would hold the altitude record." He refused to tell.)

In my opinion, the greatest achievement of flight was not Wilbur and Orville's aeroplane, nor Chuck Yeager's breaking of the sound barrier, nor the Apollo 11 lunar landing. For me, it was the launch of Voyager 2, which ballistically toured the solar system's outer planets. During the flybys, the spacecraft's slingshot trajectories stole a little of Jupiter's and Saturn's orbital energy to enable its rapid exit from the solar system. Upon passing Jupiter in 1979, Voyager's speed exceeded forty thousand miles an hour, sufficient to escape the gravitational attraction of even the Sun. Voyager passed the orbit of Pluto in 1993 and has now entered the realm of interstellar space. Nobody happens to be onboard the craft, but a gold phonograph record attached to its side is etched with the earthly sounds of, among many things, the human heartbeat. So with our heart, if not our soul, we fly ever farther.

CHAPTER FOURTEEN

GOING BALLISTIC*

In nearly all sports that use balls, the balls go ballistic at one time or another. Whether you're playing baseball, cricket, football, golf, jai alai, soccer, tennis, or water polo, a ball gets thrown, smacked, or kicked and then briefly becomes airborne before returning to Earth.

Air resistance affects the trajectories of all these balls, but regardless of what set them in motion or where they might land, their basic path is described by a simple equation found in Isaac Newton's *Principia,* his seminal 1687 book on motion and gravity. Some years later, Newton interpreted his discoveries for the Latin-literate lay reader in *The System of the World,* which includes a description of what would happen if you hurled stones horizontally at higher and higher speeds. Newton first notes the obvious: the stones would hit the ground farther and farther away from the release point, eventually landing beyond the horizon. He then reasons that if the speed were high enough, a stone would travel Earth's entire circumference, never hit the ground, and return to whack you in the back of the head. If you ducked at that instant, the object would continue forever in what is commonly called an orbit. You can't get more ballistic than that.

The speed needed to achieve low Earth orbit (affectionately called LEO) is just over seventeen thousand miles per hour—sideways—making the round trip about an hour and a half. Had Sputnik 1, the first artificial satellite, and Yuri Gagarin, the first human to travel beyond our atmo-

---

* Adapted from "Going Ballistic," *Natural History*, November 2002.

sphere, not reached that speed, they would simply have fallen back to Earth.

Newton also showed that the gravity exerted by any spherical object acts as though the object's entire mass were concentrated at its center. As a consequence, anything tossed between two people on Earth's surface is also in orbit—except that the trajectory happens to intersect the ground. This was as true for Alan B. Shepard's fifteen-minute ride aboard the Mercury spacecraft Freedom 7 in 1961 as it is for a golf drive by Tiger Woods, a home run by Alex Rodriguez, and a ball tossed by a child: they have executed what are sensibly called suborbital trajectories. Were Earth's surface not in the way, all these objects would execute perfect, albeit elongated, orbits around Earth's center. And although the law of gravity doesn't distinguish among these trajectories, NASA does. Shepard's journey was mostly free of air resistance, because it reached an altitude where there's hardly any atmosphere. For this reason alone, the media promptly crowned him America's first space traveler.

Suborbital paths are the trajectories of choice for ballistic missiles. Like a hand grenade that arcs toward its target after being hurled, a ballistic missile "flies" only under the action of gravity after being launched. These weapons of mass destruction travel hypersonically, fast enough to traverse half of Earth's circumference in forty-five minutes before plunging back to the surface at thousands of miles an hour. If a ballistic missile is heavy enough, the thing can do more damage just by falling out of the sky than can the explosion of the conventional bomb it carries.

The world's first ballistic missile was the Nazis' V-2 rocket, designed by German scientists under the leadership of Wernher von Braun. As the first object to be launched above Earth's atmosphere, the bullet-shaped, large-finned V-2 (the "V" stands for *Vergeltungswaffen,* or "Vengeance Weapon") inspired an entire generation of spaceship illustrations. After surrendering to the Allied forces, von Braun was brought to the United States, where in 1958 he directed the launch of the first US satellite. Shortly thereafter, he was transferred to the newly created National Aeronautics and Space Administration, where he developed the rocket that made America's Moon landing possible.

• • •

While hundreds of artificial satellites orbit Earth, Earth itself orbits the Sun. In his 1543 magnum opus, *De Revolutionibus,* Nicolaus Copernicus placed the Sun in the center of the known universe and asserted that Earth plus the five known planets—Mercury, Venus, Mars, Jupiter, and Saturn—executed perfect circular orbits around it. Unknown to Copernicus, a circle is an extremely rare shape for an orbit and does not describe the path of any planet in our solar system. The actual shape was deduced by German mathematician and astronomer Johannes Kepler, who published his calculations in 1609. The first of his laws of planetary motion asserts that planets orbit the Sun in ellipses.

An ellipse is a flattened circle, and the degree of flatness is indicated by a numerical quantity called eccentricity, abbreviated *e.* If *e* equals zero, you get a perfect circle. As *e* increases from zero to one, your ellipse gets more and more elongated. Of course, the greater your eccentricity, the more likely you are to cross somebody else's orbit. Comets that plunge toward Earth from the outer solar system have highly eccentric orbits, whereas the orbits of Earth and Venus closely resemble circles, with very low eccentricities. The most eccentric "planet" (now officially a dwarf planet) is Pluto, and sure enough, every time it goes around the Sun, it crosses the orbit of Neptune, behaving suspiciously like a comet.

Space Tweet #15

When asked why planets orbit in ellipses & not some other shape, Newton had to invent calculus to give an answer

May 14, 2010 3:23 AM

The most extreme example of an elongated orbit is the famous case of the hole dug all the way to China. Contrary to the expectations of our geographically challenged fellow Americans, China is not opposite the United States on the globe. The southern Indian Ocean is. To avoid emerging under two miles of water, we should dig from Shelby, Montana, to the isolated Kerguelen Islands.

Now comes the fun part.

Jump in. You now accelerate continuously in a weightless, free-fall state until you reach Earth's center—where you vaporize in the fierce heat of the iron core. Ignoring that complication, you zoom right past the center, where the force of gravity is zero, and steadily decelerate until you just reach the other side, by which time you have slowed to zero velocity. Unless a Kerguelenian instantly grabs you, you now fall back down the hole and repeat the journey indefinitely. Besides making bungee jumpers jealous, you have executed a genuine orbit, taking an hour and a half—about the same amount of time as the International Space Station.

Some orbits are so eccentric that they never loop back around again. At an eccentricity of exactly one, you have a parabola; for eccentricities greater than one, the orbit traces a hyperbola. To picture these shapes, aim a flashlight directly at a nearby wall. The emergent cone of light will form a circle. Now gradually angle the flashlight upward, and your circle distorts into ellipses of higher and higher eccentricities. When your light cone points straight up, any light that still falls on the nearby wall takes the exact shape of a parabola. Tip the flashlight away from the wall a bit more, and you've made a hyperbola. (Now you have something different to do when you go camping.) Any object with a parabolic or hyperbolic trajectory moves so fast that it will never return. If astronomers ever discover a comet with such an orbit, we will know that it has emerged from the depths of interstellar space and is on a one-time tour through the inner solar system.

Newtonian gravity describes the force of attraction between any two objects anywhere in the universe, no matter where they are found, no matter what they are made of, and no matter how large or small they may be. For example, you can use Newton's law to calculate the past and future behavior of the Earth–Moon system. But add a third object—a third source of gravity—and you severely complicate the system's motions. More generally known as the three-body problem, this ménage à trois yields richly varied trajectories whose tracking usually requires a computer.

Some clever solutions to this problem deserve attention. In one case, called the restricted three-body problem, you simplify things by assuming the third body has so little mass compared with the other two that you can ignore its presence in the equations. With this approximation, you can reliably follow the motions of all three objects in the system. And we're not cheating. Many cases like this exist in the real universe—the Sun, Jupiter, and one of Jupiter's itty-bitty moons, for instance. In another case drawn from the solar system, an entire family of rocks moves around the Sun a half-billion miles ahead of and behind Jupiter but in the same path. These are the Trojan asteroids, each one locked in its stable orbit by the gravity of Jupiter and the Sun.

Another special case of the three-body problem was discovered in recent years. Take three objects of identical mass and have them follow each other in tandem, tracing a figure eight in space. Unlike those automobile racetracks where people go to watch cars smashing into each other at the intersection of two ovals, this setup takes better care of its participants. The forces of gravity require that the system "balance" for all time at the point of intersection, and, unlike the complicated general three-body problem, all motion occurs in one plane. Alas, this special case is so odd and so rare that there is probably not a single example of it among the hundreds of billions of stars in our galaxy, and perhaps a few examples in the entire universe, making the figure-eight three-body orbit an astrophysically irrelevant mathematical curiosity.

Beyond one or two other well-behaved cases, the mutual gravity of three or more objects eventually makes their trajectories go bananas. To picture how this happens, position several objects in space. Then nudge each object according to the force of attraction between it and every other object. Recalculate all forces for the new separations. Then repeat. The exercise is not simply academic. The entire solar system is a many-body problem, with asteroids, moons, planets, and the Sun in a state of continuous mutual attraction. Newton worried greatly about this problem, which he could not solve with pen and paper. Fearing the entire solar system was unstable and would eventually crash its planets into the Sun or fling them into interstellar space, he postulated that God might step in every now and then to set things right.

The eighteenth-century French astronomer and mathematician Pierre-Simon de Laplace presented a solution to the many-body problem of the solar system more than a century later in his treatise *Mécanique Céleste.* But to do so, he had to develop a new form of mathematics known as perturbation theory. The analysis begins by assuming that there is only one major source of gravity and that all the other forces are minor yet persistent—exactly the situation that prevails in our solar system. Laplace then demonstrates analytically that the solar system is indeed stable and that you don't need new laws of physics to show this.

But how stable is it? Modern analysis demonstrates that on timescales of hundreds of millions of years—periods much longer than the ones considered by Laplace—planetary orbits are chaotic. That leaves Mercury vulnerable to falling into the Sun, and Pluto vulnerable to getting flung out of the solar system altogether. Worse yet, the solar system might have been born with dozens more planets, most of them now long lost to interstellar space. And it all started with Copernicus's simple circles.

Space Tweet #16

Trajectories unstable for 2-star systems. Must orbit far from both. Fools planet to think it orbits just 1-star

Jul 14, 2010 6:03 AM

If you could somehow rise above the plane of our galaxy, you would see each star in our Sun's neighborhood moving to and fro at ten to twenty kilometers a second. Collectively, however, those stars orbit the galaxy in wide, nearly circular paths, at speeds in excess of two hundred kilometers a second. Most of the hundreds of billions of stars in the Milky Way lie within a broad, flat disk, and—like the orbiting objects in all other spiral galaxies—the clouds, stars, and other constituents of the Milky Way thrive on big, round orbits.

If you continue to rise above the plane of the Milky Way, you would see the beautiful Andromeda galaxy, two and a half million light-years away. It's the spiral galaxy closest to us, and all the currently available data suggest we're on a collision course, plunging ever deeper into each other's gravitational embrace. Someday we will be a twisted wreck of strewn stars and colliding gas clouds. Just wait six or seven billion years. With better measurements of our relative motions, astronomers may discover a strong sideways component in addition to the motion that brings us together. If so, the Milky Way and Andromeda will instead swing past each other in an elongated orbital dance.

Whenever you're going ballistic, you're in free fall. Each of the stones whose trajectory Newton illustrated was in free fall toward Earth. The one that achieved orbit was also in free fall toward Earth, but our planet's surface curved out from under it at exactly the same rate as it fell—a consequence of the stone's extraordinary sideways motion. The International Space Station is also in free fall toward Earth. So is the Moon. And, like Newton's stones, they all maintain a prodigious sideways motion that prevents them from crashing to the ground.

A fascinating feature of free fall is the persistent state of weightlessness aboard any craft with such a trajectory. In free fall, you and everything around you fall at exactly the same rate. A scale placed between your feet and the floor would also be in free fall. Because nothing is squeezing the scale, it would read zero. For this reason, and no other, astronauts are weightless in space.

But the moment the spacecraft speeds up or begins to rotate or undergoes resistance from Earth's atmosphere, the free-fall state ends and the astronauts weigh something again. Every science-fiction fan knows that if you rotate your spacecraft at just the right speed, or accelerate your spaceship at the same rate as an object falls to Earth, you will weigh exactly what you weigh on your doctor's scale. Thus, during those long, boring journeys, you can always, in principle, simulate Earth gravity.

Another notable application of Newton's orbital mechanics is the slingshot effect. Space agencies often launch probes from Earth that have too little energy to reach their planetary destinations. Instead, the orbital wizards aim the probes along cunning trajectories that swing near a moving source of gravity, such as Jupiter. By falling toward Jupiter in the same direction as Jupiter moves, a probe can gain as much speed as the orbital speed of Jupiter itself, and then sling forward like a jai alai ball. If the planetary alignments are right, the probe can repeat the feat as it swings by Saturn, Uranus, or Neptune in turn, stealing more energy with each close encounter. Even a one-time shot at Jupiter can double a probe's speed through the solar system.

Down at the other end of the mass spectrum, there are creative ways to entertain yourself. I've always wanted to live where gravity is so weak that you could throw baseballs into orbit and effectively play catch with yourself. It wouldn't be hard. No matter how slow you pitch, there's an asteroid somewhere in the solar system with just the right gravity for you to accomplish this feat. Throw with caution, though. If you throw too fast, e could reach 1, and you'd lose the ball forever.

••• CHAPTER FIFTEEN

RACE TO SPACE*

One floodlit midnight in early October 1957, beside the river Syr Darya in the Republic of Kazakhstan—while office workers in New York were taking their afternoon break—Soviet rocket scientists were launching a two-foot-wide, polished aluminum sphere into Earth orbit. By the time New Yorkers sat down to dinner, the sphere had completed its second full orbit, and the Soviets had informed Washington of their triumph: Sputnik 1, humanity's first artificial satellite, was tracing an ellipse around Earth every ninety-six minutes, reaching a peak altitude of nearly six hundred miles.

The next morning, October 5, a report of the satellite's ascent appeared in *Pravda,* the ruling Communist Party's official newspaper. ("Sputnik," by the way, loosely translates to "fellow traveler.") Following a few paragraphs of straight facts, *Pravda* adopted a celebratory tone, ending on a note of undiluted propaganda:

> The successful launching of the first man-made earth satellite makes a most important contribution to the treasure-house of world science and culture. . . . Artificial earth satellites will pave the way to interplanetary travel and apparently our contemporaries will witness how the freed and conscientious labor of the people of the new socialist society makes the most daring dreams of mankind a reality.

---

* Adapted from "Fellow Traveler," *Natural History,* October 2007.

The space race between Uncle Sam and the Reds had begun. Round one had ended in a knockout. Ham radio operators could track the satellite's persistent beeps at 20.005 megacycles and vouch for its existence. Bird-watchers and stargazers alike—if they knew when and where to look—could see the shiny little ball with their binoculars.

And that was only the beginning: the Soviet Union won not only round one but nearly all the other rounds as well. Yes, in 1969 America put the first man on the Moon. But let's curb our enthusiasm and look at the Soviet Union's achievements during the first three decades of the Space Age.

Besides launching the first artificial satellite, the Soviets sent the first animal into orbit (Laika, a stray dog), the first human being (Yuri Gagarin, a military pilot), the first woman (Valentina Tereshkova, a parachutist), and the first black person (Arnaldo Tamayo-Méndez, a Cuban military pilot). The Soviets sent the first multiperson crew and the first international crew into orbit. They made the first spacewalk, launched the first space station, and were the first to put a manned space station into long-term orbit.

Space Tweets #17 & #18

April 12, 2011: 50 yrs ago, Yuri Gagarin is launched into orbit by Soviets. He's the 4th mammal species to achieve this feat

Apr 12, 2011 10:04 AM

Just an FYI: First mammals to achieve orbit, in order: Dog, Guinea Pig, Mouse, Russian Human, Chimpanzee, American Human

Apr 12, 2011 10:20 AM

They were also the first to orbit the Moon, the first to land an unmanned capsule on the Moon, the first to photograph Earthrise from the Moon, the first to photograph the far side of the Moon, the first to put a rover on the Moon, and the first to put a satellite in orbit around the Moon. They were the first to land on Mars and the first to land on Venus. And whereas Sputnik 1 weighed 184 pounds and Sputnik 2 (launched a month later) weighed 1,120 pounds, the first satellite America had planned to send aloft weighed slightly more than three pounds. Most ignominious of all, when the United States tried its first actual launch after Sputnik—in early

December 1957—the rocket burst into flames at the (suborbital) altitude of three feet.

In July 1955, from a podium at the White House, President Eisenhower's press secretary had announced America's intention to send "small" satellites into orbit during the International Geophysical Year (July 1957 through December 1958). A few days later a similar announcement came from the chairman of the Soviet space commission, who maintained that the first satellites shouldn't have to be all that small and that the USSR would send up a few of its own in the "near future."

And so it did.

In January 1957, the Soviet missile maven and ultrapersuasive space advocate Sergei Korolev (never referred to in the Soviet press by name) warned his government that America had declared its rockets to be capable of flying "higher and farther than all the rockets in the world," and that "the USA is preparing in the nearest months a new attempt to launch an artificial Earth satellite and is willing to pay any price to achieve this priority." His warning worked. In the spring of 1957, the Soviets began testing precursors to orbiting satellites: intercontinental ballistic missiles that could loft a two-hundred-pound payload.

On August 21, their fourth try, they succeeded. Missile and payload made it all the way from Kazakhstan to Kamchatka—some four thousand miles. TASS, the official Soviet news agency, uncharacteristically announced the event to the world:

> A few days ago a super-long-range, intercontinental multistage ballistic missile was launched. . . . The flight of the missile took place at a very great, hitherto unattained, altitude. Covering an enormous distance in a short time, the missile hit the assigned region. The results obtained show that there is the possibility of launching missiles into any region of the terrestrial globe.

Strong words. Strong motives. Enough to spook any adversary into action.

Meanwhile, in mid-July the British weekly *New Scientist* had informed its readers about the Soviet Union's growing primacy in the space race. It had even published the orbit of an impending Soviet satellite. But America took little notice.

In mid-September Korolev told an assembly of scientists about the imminent launches of both Soviet and American "artificial satellites of the Earth with scientific goals." Still America took little notice.

Then came October 4.

Sputnik 1 kicked many heads out of the sand. Some people in power went, well, ballistic. Lyndon B. Johnson, at the time the Senate majority leader, warned, "Soon [the Soviets] will be dropping bombs on us from space like kids dropping rocks onto cars from freeway overpasses." Others were anxious to downplay both the geopolitical implications of the satellite and the capabilities of the USSR. Secretary of State John Foster Dulles wrote that the importance of Sputnik 1 "should not be exaggerated" and rationalized America's nonperformance thus: "Despotic societies which can command the activities and resources of all their people can often produce spectacular accomplishments. These, however, do not prove that freedom is not the best way."

On October 5, under a page-one banner headline (and alongside coverage of a flu epidemic in New York City and the showdown in Little Rock with the segregationist Arkansas governor, Orval Faubus), the *New York Times* ran an article that included the following reassurances:

> Military experts have said that the satellites would have no practicable military application in the foreseeable future. . . . Their real significance would be in providing scientists with important new information concerning the nature of the sun, cosmic radiation, solar radio interference and static-producing phenomena.

What? No military applications? Satellites were simply about monitoring the Sun? Behind-the-scenes strategists thought otherwise. According to the summary of an October 10 meeting between President Eisenhower and his National Security Council, the United States had "always been aware of the cold war implications of the launching of the first earth satellite." Even

America's best allies "require assurance that we have not been surpassed scientifically and militarily by the USSR."

Eisenhower didn't have to worry about ordinary Americans, though. Most remained unperturbed. Or maybe the spin campaign worked its magic. In any case, plenty of ham radio operators ignored the beeps, plenty of newspapers ran their satellite articles on page three or five, and a Gallup poll found that 60 percent of people questioned in Washington and Chicago expected that the United States would make the next big splash in space.

America's cold warriors, now fully awake to the military potential of space, understood that US postwar prestige and power had been challenged. Within a year, money to help restore them would be pumped into science education, the education of college teachers, and research useful to the military.

Back in 1947, the President's Commission on Higher Education had proposed as a goal that a third of America's youth should graduate from a four-year college. The National Defense Education Act of 1958 was a key, if modest, push in that direction. It provided low-interest student loans for undergraduates as well as three-year National Defense Fellowships for several thousand graduate students. Funding for the National Science Foundation tripled right after Sputnik; by 1968 it was a dozen times the pre-Sputnik appropriation. The National Aeronautics and Space Act of 1958 hatched a new, full-service civilian agency called the National Aeronautics and Space Administration—NASA. The Defense Advanced Research Projects Agency, or DARPA, was born the same year.

All those initiatives and agencies funneled the best American students into science, math, and engineering. The government got a lot of bang for its buck; graduate students in those fields, come wartime, got draft deferments; and the concept of federal funding for education got validated.

But some kind of satellite, built by any means necessary, had to be launched ASAP. Luckily, during the closing weeks and immediate aftermath of World War II in Europe, the United States had acquired a worthy challenger to Sergei Korolev: the German engineer and physicist Wernher von Braun, former leader of the team that had developed the terrifying V-2 ballistic missile. We also acquired more than a hundred members of his team.

Instead of being put on trial at Nuremburg for war crimes, von Braun became America's savior, the progenitor and public face of the US space program. His first high-profile task was to provide the first rocket for the first successful launch of America's first satellite. On January 31, 1958—less than four months after Sputnik 1's round-the-world tour—he and his rocketeers got the thirty-pound Explorer 1, plus its eighteen pounds of scientific instrumentation, into orbit.

Space Tweet #19

An object in orbit has high sideways speed so it falls to Earth at exactly the same rate that the round Earth curves below it

May 14, 2010 11:56 AM

Disposal of dead weight was a key to their success. If you want to reach orbital speeds—just over seventeen thousand miles an hour—you'd better unladen your rocket at every opportunity. Rocket motors are heavy, fuel tanks are heavy, fuel itself is heavy, and every kilogram of unnecessary mass schlepped into space wastes thousands of kilograms of fuel. The solution? The multistage rocket. When the first-stage fuel tank is spent, throw it away. Run out of fuel in the next stage; throw that away too.

Jupiter-C, the rocket that launched Explorer 1, weighed 64,000 pounds at takeoff, fully loaded. The final stage weighed 80.

Like the R-7 rocket that launched Sputnik 1, the Jupiter-C was a modified weapon. The science was a secondary, even tertiary, outgrowth of military R&D. Cold warriors wanted bigger and more lethal ballistic missiles, with nuclear warheads crammed into the nose cones.

High ground is the military's best friend, and what ground could be higher than a satellite orbiting no more than forty-five minutes away from a possible target? Thanks to Sputnik 1 and its successors, the USSR held that high ground until 1969, when, courtesy of von Braun and colleagues, the USA's Saturn V rocket took the Apollo 11 astronauts to the Moon.

Today, whether Americans know it or not, a new space race is under way. This time, America faces not only Russia but also China, the European Union, India, and more. Maybe this time the race will be one between fellow travelers rather than potential adversaries—more about fostering innovations in science and technology than about struggling to rule the high ground.

••• CHAPTER SIXTEEN

2001—FACT VS. FICTION*

The long-awaited year has come and gone. There was no escape from the relentless comparisons between the spacefaring future we saw in Stanley Kubrick's *2001: A Space Odyssey* and the reality of our measly earthbound life in the real 2001. We don't yet have a lunar base camp, and we have not yet sent hibernating astronauts to Jupiter in outsize spaceships, but we have nonetheless come a long way in our exploration of space.

Today, the greatest challenge to human exploration of space, apart from money and other political factors, is surviving biologically hostile environments. We need to send into space an improved version of ourselves—doppelgangers who can somehow withstand the extremes of temperature, the high-energy radiation, and the meager air supply, yet still conduct a full round of scientific experiments.

Fortunately, we have already invented such things: they're space robots. They don't look humanoid and we don't refer to them as "who," but they conduct all of our interplanetary exploration. You don't have to feed them, they don't need life support, and they won't get upset if you don't bring them home. Our ensemble of space robots includes probes that are monitoring the sun, orbiting Mars, intercepting a comet's tail, orbiting an asteroid, orbiting Saturn, and heading to Jupiter and Pluto.

Four of our early space probes were launched with enough energy and

---

* Adapted from "2001, for Real," Op-Ed, *The New York Times*, January 1, 2001.

with the right trajectory to escape the solar system altogether, each one carrying encoded information about humans for the intelligent aliens who might recover the hardware.

Even though humans have not left footprints on Mars or on Jupiter's moon Europa, our space robots at these worlds have beamed back to us compelling evidence of the presence of water. These discoveries fire our imaginations with the prospect of finding life on future missions.

We also maintain hundreds of communication satellites, as well as a dozen space-based telescopes that see the universe in different bands of light, including infrared and gamma rays. In particular, the microwave band allows us to see the edge of the observable universe, where we find evidence of the Big Bang.

And so, we may have no interplanetary colonies or other unrealized dreamscapes, but our presence in space has been growing exponentially nonetheless. In some ways, space exploration in the real 2001 strongly resembles that of Kubrick's movie. Apart from our flock of robotic probes, we have a fleet of hardware in the sky. Just as they do in *2001* the movie, we've got a space station. It was assembled with parts delivered by reusable, docking space shuttles (which happened to say "NASA" on the side instead of "Pan Am"). And, as in the movie, the space station has zero-G flush toilets, with complicated instructions, and plastic pouches of unappealing astronaut food.

As far as I can tell, the only things Kubrick's movie has that we don't have are Johann Strauss's "Blue Danube" waltz filling the vacuum of space, and a homicidal mainframe named HAL.

••• CHAPTER SEVENTEEN

LAUNCHING THE RIGHT STUFF*

In 2003 the space shuttle orbiter Columbia broke into pieces over central Texas. A year later, President George W. Bush announced a long-term program of space exploration that would return humans to the Moon and thereafter send them to Mars and beyond. Over that time, and for years to come, the twin Mars Exploration Rovers, Spirit and Opportunity, wowed scientists and engineers at the rovers' birthplace—NASA's Jet Propulsion Laboratory (JPL)—with their skills as robotic field geologists.

The confluence of these and other events resurrects a perennial debate: with two failures out of 135 shuttle missions during the life of the manned space program, and its astronomical expense relative to robotic programs, can sending people into space be justified, or should robots do the job alone? Or, given society's sociopolitical ailments, is space exploration something we simply cannot afford to pursue? As an astrophysicist, as an educator, and as a citizen, I'm compelled to speak my mind on these issues.

Modern societies have been sending robots into space since 1957, and people since 1961. Fact is, it's vastly cheaper to send robots—in most cases, a fiftieth the cost of sending people. Robots don't much care how hot or cold space gets; give them the right lubricants, and they'll operate in a vast range of temperatures. They don't need elaborate life-support systems either. Robots can spend long periods of time moving around and among the planets, more or less unfazed by ionizing radiation. They do not lose

* Adapted from "Launching the Right Stuff," *Natural History*, April 2004.

bone mass from prolonged exposure to weightlessness, because, of course, they are boneless. Nor do they have hygiene needs. You don't even have to feed them. Best of all, once they've finished their jobs, they won't complain if you don't bring them home.

So if my only goal in space is to do science, and I'm thinking strictly in terms of the scientific return on my dollar, I can think of no justification for sending a person into space. I'd rather send the fifty robots.

But there's a flip side to this argument. Unlike even the most talented modern robots, humans are endowed with the ability to make serendipitous discoveries that arise from a lifetime of experience. Until the day arrives when bioneurophysiological computer engineers can do a human-brain download on a robot, the most we can expect of the robot is to look for what it has already been programmed to find. A robot—which is, after all, a machine for embedding human expectations in hardware and software—cannot fully embrace revolutionary scientific discoveries. And those are the ones you don't want to miss.

In the old days, people generally pictured robots as a hunk of hardware with a head, neck, torso, arms, and legs—and maybe some wheels to roll around on. They could be talked to and would talk back (sounding, of course, robotic). The standard robot looked more or less like a person. The fussbudget character C3PO, from the *Star Wars* movies, is a perfect example.

Even when a robot doesn't look humanoid, its handlers might present it to the public as a quasi-living thing. Each of NASA's twin Mars rovers, for instance, was described in JPL press packets as having "a body, brains, a 'neck and head,' eyes and other 'senses,' an arm, 'legs,' and antennas for 'speaking' and 'listening.' " On February 5, 2004, according to the status reports, "Spirit woke up earlier than normal today . . . in order to prepare for its memory 'surgery.' " On the 19th the rover remotely examined the rim and surrounding soil of a crater dubbed Bonneville, and "after all this work, Spirit took a break with a nap lasting slightly more than an hour."

In spite of all this anthropomorphism, it's pretty clear that a robot can have any shape at all: it's simply an automated piece of machinery that accomplishes a task, either by repeating an action faster or more reliably than the average person can, or by performing an action that a person, relying solely on the five senses, would be unable to accomplish. Robots that paint cars on assembly lines don't look much like people. The Mars rovers looked a bit like toy flatbed trucks, but they could grind a pit in the surface of a rock, mobilize a combination microscope-camera to examine the freshly exposed surface, and determine the rock's chemical composition—just as a geologist might do in a laboratory on Earth.

It's worth noting, by the way, that even a human geologist doesn't go it alone. Unaided by some kind of equipment, a person cannot grind down the surface of a rock; that's why a field geologist carries a hammer. To analyze a rock further, the geologist deploys another kind of apparatus, one that can determine its chemical composition. Therein lies a conundrum. Almost all the science likely to be done in an alien environment would be done by some piece of equipment. Field geologists on Mars would lug it around on their daily strolls across a Martian crater or outcrop, where they might take measurements of the soil, the rocks, the terrain, and the atmosphere. But if you can get a robot to haul and deploy all the same instruments, why send a field geologist to Mars at all?

One good reason is the geologist's common sense. Each Mars rover was designed to move for about ten seconds, then stop and assess its immediate surroundings for twenty seconds, then move for another ten seconds, and so on. If the rover moved any faster, or moved without stopping, it might stumble on a rock and tip over, becoming as helpless as a Galápagos tortoise on its back. In contrast, a human explorer would just stride ahead, because people are quite good at watching out for rocks and cliffs.

Back in the late 1960s and early 1970s, in the days of NASA's manned Apollo flights to the Moon, no robot could decide which pebbles to pick up and bring home. But when the Apollo 17 astronaut Harrison Schmitt, the only geologist (in fact, the only scientist) to have walked on the Moon, noticed some odd orange soil on the lunar surface, he immediately collected a sample. It turned out to be minute beads of volcanic glass. Today a robot can perform staggering chemical analyses and transmit amazingly detailed images, but it still can't react efficiently, as Schmitt did, to a surprise. By contrast, packed inside the field geologist are the capacities to walk, run, dig, hammer, see, communicate, interpret, and invent.

Of course when something goes wrong, an on-the-spot human being becomes a robot's best friend. Give a person a wrench, a hammer, and some duct tape, and you'd be surprised what can get fixed. After landing on Mars, did the Spirit rover just roll right off its platform and start checking out the neighborhood? No, its airbags were blocking the path. Not until twelve more days had passed did Spirit's remote controllers manage to get all six of its wheels rolling on Martian soil. Anyone on the scene on January 3 could have just lifted the airbags out of the way and in mere seconds given Spirit a little shove.

Let's assume, then, that we can agree on a few things: People notice the unexpected, react to unforeseen circumstances, and solve problems in ways that robots cannot. Robots are cheap to send into space but can make only a preprogrammed analysis. Cost and scientific results, however, are not the only relevant issues. There's also the question of exploration.

The first troglodytes to cross the valley or climb the mountain ventured forth from the family cave not because they wanted to make a scientific discovery but because something unknown lay beyond the horizon. Perhaps they sought more food, better shelter, or a more promising way of life. In any case, they felt the urge to explore. It may be hardwired, lying deep within the behavioral identity of the human species. How else could our ancestors have migrated from Africa to Europe and Asia, and onward to North and South America? To send a person to Mars who can look under the rocks or find out what's down in the valley is the natural extension of what ordinary people have always done on Earth.

Many of my colleagues assert that plenty of science can be done without putting people in space. But if they were kids in the 1960s, and you ask what inspired them to become scientists, nearly every one (at least in my experience) will cite the high-profile Apollo program. It took place when they were young, and it's what got them excited. Period. In contrast, even if they also mention the launch of Sputnik 1, which gave birth to the space era, very few of those scientists credit their interest to the numerous other unmanned satellites and space probes launched by both the United States and the Soviet Union shortly after Sputnik.

So if you're a first-rate scientist drawn to the space program because you'd initially been inspired by astronauts rocketing into the great beyond, it's somewhat disingenuous of you to contend that people should no longer go into space. To take that position is, in effect, to deny the next generation of students the thrill of following the same path you did: enabling one of our own kind, not just a robotic emissary, to walk on the frontier of exploration.

Whenever we hold an event at the Hayden Planetarium that includes an astronaut, I've found there's a significant uptick in attendance. Any astronaut will do, even one most people have never heard of. The one-on-one encounter makes a difference in the hearts and minds of Earth's armchair space travelers—whether retired science teachers, hardworking bus drivers, thirteen-year-old kids, or ambitious parents.

Of course, people can and do get excited about robots. From January 3 through January 5, 2004, the NASA website that tracked the doings of the Mars rovers sustained more than half a billion hits—506,621,916 to be exact. That was a record for NASA, surpassing the world's web traffic in pornography over the same three days.

The solution to the quandary seems obvious to me: send both robots and people into space. Space exploration needn't be an either/or transaction, because there's no avoiding the fact that robots are better suited for certain tasks, and people for others.

One thing is certain: in the coming decades, the United States will need to call upon multitudes of scientists and engineers from scores of disciplines, and astronauts will need to be extraordinarily well trained. The search for evidence of past life on Mars, for instance, will require top-notch biologists. But what does a biologist know about planetary terrains? Geologists and geophysicists will have to go too. Chemists will be needed to check out the atmosphere and test the soils. If life once thrived on Mars, the remains might now be fossilized, and so perhaps we'll need a few paleontologists to join the fray. People who know how to drill through kilometers of soil and rock will also be must-haves, because that's where Martian water reserves might be hiding.

Where will all those talented scientists and technologists come from? Who's going to recruit them? Personally, when I give talks to students old enough to decide what they want to be when they grow up but young enough not to get derailed by raging hormones, I need to offer them a tasty carrot to get them excited enough to become scientists. That task is made easy if I can introduce them to astronauts in search of the next generation to share their grand vision of exploration and join them in space. Without such inspiring forces behind me, I'm just that day's entertainment. My reading of history and culture tells me that people need their heroes.

Twentieth-century America owed much of its security and economic strength to its support for science and technology. Some of the most revolutionary (and marketable) technology of past decades has been spun off the research done under the banner of US space exploration: kidney dialysis machines, implantable pacemakers, LASIK surgery, global positioning satellites, corrosion-resistant coatings for bridges and monuments (including the Statue of Liberty), hydroponic systems for growing plants, collision-avoidance systems on aircraft, digital imaging, infrared handheld cameras, cordless power tools, athletic shoes, scratch-resistant sunglasses, virtual reality. And that list doesn't even include Tang.

Although solutions to a problem are often the fruit of direct investment in targeted research, the most revolutionary solutions tend to emerge from cross-pollination with other disciplines. Medical investigators might never have known of X-rays, since they do not naturally occur in biological systems. It took a physicist, Wilhelm Conrad Röntgen, to discover these light rays that could probe the body's interior with nary a cut from a surgeon.

Here's another example of cross-pollination. Soon after the Hubble Space Telescope was launched in April 1990, NASA engineers realized that the telescope's primary mirror—which gathers and reflects the light from celestial objects into its cameras and spectrographs—had been ground to an incorrect shape. In other words, the two-billion-dollar telescope was producing fuzzy images.

That was bad.

As if to make lemonade out of lemons, though, computer algorithms came to the rescue. Investigators at the Space Telescope Science Institute in Baltimore, Maryland, developed a range of clever and innovative image-processing techniques to compensate for some of Hubble's shortcomings.

Turns out, maximizing the amount of information that could be extracted from a blurry astronomical image is technically identical to maximizing the amount of information that can be extracted from a mammogram. Soon the new techniques came into common use for detecting early signs of breast cancer.

But that's only part of the story.

In 1997, for Hubble's second servicing mission (the first, in 1993, corrected the faulty optics), shuttle astronauts swapped in a brand-new, high-resolution digital detector—designed to the demanding specs of astrophysicists whose careers are based on being able to see small, dim things in the cosmos. That technology is now incorporated in a minimally invasive, low-cost system for doing breast biopsies, the next stage after mammograms in the early diagnosis of cancer.

So why not ask investigators to take direct aim at the challenge of detecting breast cancer? Why should innovations in medicine have to wait for a Hubble-size blunder in space? My answer may not be politically correct, but it's the truth: when you organize extraordinary missions, you attract people of extraordinary talent who might not have been inspired by or attracted to the goal of saving the world from cancer or hunger or pestilence.

Today, cross-pollination between science and society comes about when you have ample funding for ambitious long-term projects. America has profited immensely from a generation of scientists and engineers who, instead of becoming lawyers or investment bankers, responded to a challenging vision posed in 1961 by President John F. Kennedy. Proclaiming the intention to land a man on the Moon, Kennedy welcomed the citizenry to aid in the effort. That generation, and the one that followed, was the same generation of technologists who invented the personal computer. Bill Gates, cofounder of Microsoft, was thirteen years old when the United States landed an astronaut on the Moon; Steve Jobs, cofounder of Apple Computer, was fourteen. The PC did not arise from the mind of a banker or artist or professional athlete. It was invented and developed by a technically trained workforce, who had responded to the dream unfurled before them and were thrilled to become scientists and engineers.

Yes, the world needs bankers and artists and even professional athletes. They, among countless others, create the breadth of society and culture. But if you want tomorrow to come—if you want to spawn entire economic sectors that didn't exist yesterday—those are not the people you turn to. It is technologists who create that kind of future. And it is visionary steps into space that create that kind of technologist. I look forward to the day when the solar system becomes our collective backyard—explored not only with robots, but with the mind, body, and soul of our species.

••• CHAPTER EIGHTEEN

THINGS ARE LOOKING UP*

On September 8, 2004, the scientific payload from NASA's Genesis mission crashed into the Utah desert at nearly two hundred miles per hour after its parachutes failed to open. The spacecraft had spent three years orbiting the Sun at a distance of nearly a million miles from Earth, collecting some of the tiny atomic nuclei that the Sun expels continuously into space, and whose abundances encode the original composition of the material from which the solar system formed 4.6 billion years ago. NASA scientists have recovered some of the results from Genesis, and thus avoided writing off their time and our $260 million as a total loss.

But even if no usable data had returned, this single failure merely emphasizes how well we are doing as we explore the cosmos. NASA's two robotic geologists roving the surface of Mars have both exceeded their scheduled lifetimes while returning stunning images of the Martian surface—images that tell us Mars once had running water and large lakes or seas. The Mars Global Surveyor, likewise operating well beyond its planned lifetime, continues to orbit the Red Planet and send us high-resolution images of the Martian surface. And the European Space Agency's Mars Express Orbiter has supplied evidence of methane in the Martian atmosphere, which may be traceable to active underground bacterial colonies. The Cassini spacecraft orbits Saturn, and Cassini's Huygens probe detached and then descended through the smoggy atmosphere of Saturn's largest

* Adapted from unpublished op-ed by Neil deGrasse Tyson and Donald Goldsmith, September 2004.

moon, Titan, landed on its surface, and confirmed the existence of liquid lakes of methane. Titan itself may well prove to be a site for life of a different kind. We also have MESSENGER, the first probe to orbit the Sun's innermost planet.

When we turn to the much vaster cosmos beyond our solar system, we find a stunning array of spacecraft that orbit Earth outside our interfering atmosphere. NASA's orbiting Chandra X-ray Observatory detects X-rays from distant venues of cosmic violence, such as the turbulent environs that surround hungry black holes, while NASA's Spitzer Space Telescope maps infrared light, a calling card of young stars and star-forming regions. The European Space Agency's Integral satellite studies gamma rays, the highest-energy form of light, which arise from exploding stars and other violent cosmic events; NASA's Swift Gamma Ray Burst Explorer searches for the most distant gamma-ray outbursts in the universe. Meanwhile, the Hubble Space Telescope will continue to work until its larger successor, the James Webb Space Telescope, reaches orbit, peering farther than any previous telescope as it chronicles the formation of galaxies and the large-scale structures they trace.

Enlightened by our surrogate eyes in this busy vacuum of space, we should occasionally remind ourselves that Earth's continents display no national boundaries. But above all else, our smallness in the vastness of the universe should humble us all.

••• CHAPTER NINETEEN

FOR THE LOVE OF HUBBLE*

The Hubble Space Telescope, the most productive scientific instrument of all time, had its fifth and final repair mission in the spring of 2009. The space shuttle astronauts launched from Kennedy Space Center in Florida, matched orbits with the telescope, captured it, serviced it, upgraded it, and replaced its broken parts—on the spot.

Roughly the size of a Greyhound bus, Hubble was launched aboard the space shuttle Discovery in 1990 and has substantially outlived its initial ten-year life expectancy. For students in high school today, Hubble has been their primary conduit to the cosmos. The final servicing mission extended Hubble's life several years. Among other things, it replaced burned-out circuit boards in the Advanced Camera for Surveys. That's the instrument responsible for Hubble's most memorable images since its installation in 2002.

Servicing Hubble requires exquisite dexterity. I recently had the opportunity to visit NASA's Goddard Space Flight Center in Maryland. There I donned puffy, pressurized astronaut gloves, wielded a space-age portable screwdriver, stuck my head in a space helmet, and attempted to extract a faulty circuit board in a mock-up of the failed camera, which was embedded within a full-scale model of the Hubble. This was a darn near impossi-

* Adapted from "For the Love of Hubble," *Parade*, June 22, 2008.

ble feat. And I wasn't weightless. I was not wearing the full-body spacesuit. Nor were Earth and space drifting by.

Normally we think of astronauts as brave and noble. But in this case, having the "right stuff" includes being a hardware surgeon.

Hubble is not alone up there. Dozens of space telescopes of assorted sizes and shapes orbit Earth and the Moon. Each one provides a view of the cosmos that is unobstructed, unblemished, and undiminished by Earth's turbulent and murky atmosphere. But most of these telescopes were launched with no means of servicing them. Parts wear out. Gyroscopes fail. Coolant evaporates. Batteries die. Hardware realities limit a telescope's life expectancy.

All these telescopes advance science, but most perform their duties without the public's awareness or adulation. They are designed to detect bands of light invisible to the human eye, some of which never penetrate Earth's atmosphere. Entire classes of objects and phenomena in the cosmos reveal themselves only through one or more invisible cosmic windows. Black holes, for example, were discovered by their X-ray calling card—radiation generated by the surrounding, swirling gas just before it descended into the abyss. Telescopes have also captured microwave radiation—the primary physical evidence for the Big Bang.

Hubble, on the other hand, is the first and only space telescope to observe the universe using primarily visible light. Its stunningly crisp, colorful, and detailed images of the cosmos make Hubble a kind of supreme version of human eyes in space. Yet its appeal derives from much more than a stream of pretty portraits. Hubble came of age in the 1990s, during exponential growth of access to the Internet. That's when its digital images were first cast into the public domain. As we all know, anything that's fun, free, and forwardable spreads rapidly online. Soon Hubble images, one more splendorous than the next, became screensavers and desktop wallpaper for computers owned by people who would never before have had the occasion to celebrate, however quietly, our place in the universe.

Indeed, Hubble brought the universe into our backyard. Or rather, it expanded our backyard to enclose the universe itself, accomplishing that with images so intellectually, visually, and even spiritually fulfilling that most don't even need captions. No matter what Hubble reveals—planets, dense star fields, colorful interstellar nebulae, deadly black holes, graceful colliding galaxies, the large-scale structure of the universe—each image establishes your own private vista on the cosmos.

Space Tweet #20

In the era of Hubble & space probes, dots of light on the night sky have become worlds. Worlds have become our backyard

Feb 20, 2011 6:56 PM

Hubble's scientific legacy is unimpeachable. More research papers have been published using its data than have ever been published for any other scientific instrument in any discipline. Among Hubble's highlights is its settling of the decades-old debate about the age of the universe. Previously, the data were so bad that astrophysicists could not agree to within a factor of two. Some thought ten billion years; others, twenty billion. Yes, it was embarrassing. But Hubble enabled us to measure accurately how the brightness varies in a particular type of distant star. That information, when plugged into a simple formula, provides that star's distance from Earth. And because the entire universe is expanding at a known rate, we can then turn back the clock to determine how long ago everything was in the same place. The answer? The universe was born 13.7 billion years ago.

Another result, long suspected to be true but confirmed by Hubble, was the discovery that every large galaxy, such as our own Milky Way, has a supermassive black hole at its center that dines on stars, gas clouds, and other unsuspecting matter that wanders too close. The centers of galaxies are so densely packed with stars that atmospherically blurred Earth-based telescopes see only a mottled cloud of light—the puddled image of hundreds or thousands of stars. From space, Hubble's sharp detectors allow us to see each star individually and to track its motion around the galactic center. Behold, these stars move much, much faster than they have any right to. A small, unseen yet powerful source of gravity must be tugging on them. Crank the equations, and we are forced to conclude that a black hole lurks in their midst.

In 2004, a year after the Columbia tragedy, NASA announced that Hubble would not receive its last servicing mission. Curiously, the loudest voices of dissent were from the general public. Akin to a modern version of a torch-wielding mob, they voiced their opposition in every medium available, from op-eds to petitions. Ultimately, Congress listened and reversed the decision. Democracy had a shining moment: Hubble would indeed be serviced one last time.

Of course, nothing lasts forever—nothing except, perhaps, the universe itself. So Hubble eventually will die. But in the meantime, the James Webb Space Telescope beckons, designed to see deeper into the universe than Hubble ever could. When launched, funding permitting, it will allow us to plumb the depths of gas clouds in our own Milky Way galaxy in search of stellar nurseries, as well as to probe the earliest epochs of the universe in search of the formation of galaxies themselves.

NASA retired the aging space shuttle in 2011. Given sufficient political will, this step should enable its aerospace engineers, assembly lines, and funding streams to focus on a new suite of launch vehicles designed to do what the shuttles can't: take us beyond low Earth orbit, with sights on farther frontiers.

••• CHAPTER TWENTY

HAPPY ANNIVERSARY, APOLLO 11*

The National Air and Space Museum is unlike any other place on this planet. If you're hosting visitors from another country and they want to know what single museum best captures what it is to be American, this is the museum you take them to. Here they can see the 1903 Wright Flyer, the 1927 Spirit of St. Louis, the 1926 Goddard rocket, and the Apollo 11 command module—silent beacons of exploration, of a few people willing to risk their lives for the sake of discovery. Without those risk takers, society rarely goes anywhere.

We celebrate the fortieth anniversary of the Moon landing, July 20, 1969. Forty: that's a big number. How many days was the Ark at sea? Forty. (Also forty nights.) How many years did Moses wander the desert? Forty.

The Apollo era stoked ambitions. Many of us are here because of it. But the struggle is not over. Not everybody was part of that vision. Not everybody was struck by it. And I blame us for that. All space people feel it. You know and understand the majestic journey. Yet there are those who don't, who haven't even thought about it. Two-thirds of the people alive today in the world were born after 1969. Two-thirds.

Do you remember Jay Leno doing his Jaywalking for NBC's *Tonight Show*? He'd go out in the street and ask people a simple question. Once he

---

* Adapted from Master of Ceremonies remarks, fortieth anniversary celebration of the Apollo 11 Moon landing, Smithsonian National Air and Space Museum, Washington, DC, July 20, 2009.

went up to a freshly minted college graduate and asked, "How many moons does Earth have?" Here's her reply: "How do you expect me to remember that? I had astronomy two semesters ago."

That scares me.

Today we have assembled many astronauts who were part of the first wave of America's space explorers—heroes of a generation. There are also heroes who never flew. And those who mattered to us as a nation who are now gone. Walter Cronkite passed away just this past Friday at the age of ninety-two. At first I was saddened when I learned of his death. But when you're that old and you die, it's not an occasion to be sad; it's an occasion to celebrate a life. Cronkite: the most trusted man in America. We all knew him as a supporter of space. He anchored the *CBS Evening News* with intelligence, integrity, and compassion.

I remember when I was a kid and I first learned there was someone by the name of Cronkite. Do you know anyone else named Cronkite, other than Walter? I don't think so. So the name was interesting to me. I knew enough about the periodic table that it sounded like a new element. You know, we have aluminum, nickel, silicon. There's the fictional kryptonite. And then there's cronkite.

One of my most indelible memories of Walter is from when I was ten years old. At 7:51 A.M. on December 21, 1968—exactly the scheduled time—Apollo 8 lifted off from Kennedy Space Center. It was the first mission ever to leave low Earth orbit, the first time anyone ever had a destination other than Earth. When Walter Cronkite announced that the Apollo 8 command module—en route to the Moon—had just left the gravitational pull of Earth, I was taken aback. How could that be? They hadn't reached the Moon yet, and of course the Moon lies within Earth's gravity. Later I would learn, of course, that he was referring to a Lagrangian point between Earth and the Moon—a point where all forces of the Earth–Moon system balance. When you cross it, you fall toward the Moon instead of back toward Earth. And so I learned a bit of physics from Walter Cronkite. Godspeed to this voice of America, who died on the fortieth anniversary of Apollo 11. What a way to go.

• • •

It's been a busy week. We lose Walter Cronkite; we gain some appointments. The United States Senate confirmed the new NASA administrator and the new deputy NASA administrator, Charles F. Bolden Jr. and Lori B. Garver. Lori Garver—her whole life has been in space. She started working for John Glenn in 1983. She was executive director of the National Space Society and president of Capital Space, LLC. I've known Lori Garver for fifteen years; I've known Charlie Bolden for fifteen minutes. Just met him in the green room. The man looks like he came from central casting: four decades in public service, a combat pilot for the Marines, fourteen years as a member of NASA's astronaut corps. The confirmation hearings began like a love fest, with senators from everywhere saying, "Charlie's the man."

As I'm sure you know, decisions at NASA don't happen in a vacuum. I've participated in two commissions in the service of NASA: the Commission on the Future of the United States Aerospace Industry (its final report, from 2002, was called *Anyone, Anything, Anywhere, Anytime*) and the President's Commission on Implementation of United States Space Exploration Policy (the final report, from 2004, was called *A Journey to Inspire, Innovate, and Discover: Moon, Mars and Beyond*). We were trying to study what is, what isn't, what should be, and what's possible. As one of the commissioners, I remember being bombarded by the public and by people from the aerospace community. Everybody has an idea about what NASA should do. Somebody's got a new design for a rocket, or a desired destination, or a new propellant. Initially I felt as though people were interfering with my getting our job done. But then I stepped back and realized that if so many people want to tell NASA what to do, it's a good sign, not a bad sign. There I was being annoyed, when in fact I should have celebrated it as an expression of love for the future of NASA.

The agency continues to solicit input from experts. A committee headed by Norm Augustine has studied the future of NASA's manned spaceflight program (the final report, from late 2009, is titled *Seeking a Human Spaceflight Program Worthy of a Great Nation*). You could go online to hsf.nasa.gov—"hsf" for human spaceflight—and tell them what you think. How many countries allow such a thing, much less suggest you might be able to influence the direction an agency will take?

As some of you know, I'm an astrophysicist—less a space person than a science person. I care about exploding stars, black holes, and the fate of the Milky Way. And not all space missions are about building a space station.

One of my favorite recent missions was when the space shuttle Atlantis serviced the Hubble Space Telescope. In May 2009, Atlantis's astronauts—I prefer to think of them as them astrosurgeons—repaired and upgraded Hubble. They conducted five spacewalks during their mission to extend the life of the telescope at least five years, possibly ten—literally a new lease on life. They successfully installed two new instruments, repaired two others, replaced gyroscopes and batteries, added new thermal insulation to protect the most celebrated telescope since the era of Galileo. It was the crowning achievement of what can happen when the manned space program is in synchrony with the robotic program.

Space Tweet #21
Space Shuttle Atlantis – final trip before retirement today. On board, a chunk from Isaac Newton's apple tree. Cool
May 14, 2010 2:22 AM

By the way, Hubble is beloved not only because it has taken such great pictures, but because it's been around a long time. No other space telescopes were designed to be serviced. You put them up; the coolant runs out after three years; the gyros go out after five; they drop in the Pacific after six. That's not enough time for the public to warm up to these instruments, to learn what they do and why.

Inspiration is manifested in many ways. Space itself is a catalyst. It operates in our hearts and our souls and our minds and our creativity. It's not just the target of a science experiment—space is embedded in our culture. In

2004 NASA announced the creation of a special honor, the Ambassador of Exploration award. It's not given out every year, nor is it given out to just anyone. The award is a small sample of the 842 pounds of rocks and soil that have come from the Moon during America's six expeditions there, and it is presented to honor the first generation of explorers and to renew our commitment to expand that enterprise.

Tonight, we are honored to present the Ambassador of Exploration award to the family of President John Fitzgerald Kennedy. Certainly most of us remember President Kennedy's speech to a special joint session of Congress in May 1961, in which he declared the goal of putting an American on the Moon within the decade. But perhaps not quite so many are familiar with the "Moon speech" he gave the following year at the Rice University stadium in Houston, Texas. Early in that speech, the president mentioned that most of the total number of scientists who had ever lived on Earth were currently alive. He then presented the sweep of history in capsule form:

> Condense, if you will, the fifty thousand years of man's recorded history in a time span of but a half century. Stated in these terms, we know very little about the first forty years, except at the end of them advanced man had learned to use the skins of animals to cover [himself]. Then about ten years ago, under this standard, man emerged from his caves to construct other kinds of shelter. Only five years ago man learned to write and use a cart with wheels. . . . The printing press came this year, and then less than two months ago, during this whole fifty-year span of human history, the steam engine provided a new source of power. . . . Last month electric lights and telephones and automobiles and airplanes became available. Only last week did we develop penicillin and television and nuclear power, and now, if America's new spacecraft succeeds in reaching Venus, we will have literally reached the stars before midnight tonight.

Repeatedly Kennedy spoke of the necessity of America's being first, being the leader, doing what is hard rather than what is easy, and he described, to an audience for whom going into space was new and breathtaking, the multiple US space endeavors that were already under way and the several US satellites that were already orbiting. He didn't hesitate to announce how much money he wanted for the space budget—"fifty cents a week for every man, woman and child in the United States, for we have given this program a high national priority"—but then justified that generous funding by presenting a vivid picture of the outcome he envisioned:

> But if I were to say, my fellow citizens, that we shall send to the Moon, 240,000 miles away from the control station in Houston, a giant rocket more than three hundred feet tall, the length of this football field, made of new metal alloys, some of which have not yet been invented, capable of standing heat and stresses several times more than have ever been experienced, fitted together with a precision better than the finest watch, carrying all the equipment needed for propulsion, guidance, control, communications, food and survival, on an untried mission, to an unknown celestial body, and then return it safely to Earth, reentering the atmosphere at speeds of over 25,000 miles per hour, causing heat about half that of the temperature of the Sun . . . and do all this, and do it right, and do it first before this decade is out—then we must be bold.

Who could remain uninspired by such words!

Neil Armstrong, commander of Apollo 11, was part of NASA long before NASA formally existed. He was a naval aviator, the youngest pilot in his squadron. He flew seventy-eight combat missions during the Korean War. Neil Armstrong is someone with firsthand experience of the Moon, someone who's had both a bird's-eye view and a moonwalker's view of the Sea of Tranquillity.

Some people seem to believe that we just strap the astronauts to a rocket and fire them to the Moon. Fact is, a lot of image reconnaissance goes into planning these journeys. For example, in 1966–67 five Lunar Orbiter spacecraft were sent to study the Moon and photograph possible landing sites. The photograph of what became Apollo 11's landing site is now part of the Lunar Orbiter Image Recovery Project at the NASA Ames Research

Center. Fast forward four decades, and the NASA's Lunar Reconnaissance Orbiter, the LRO, returned its first images of the Apollo 11 landing site, with the descent stage of the lunar module still sitting right there, casting a long, distinctive shadow. LRO is the next step in returning astronauts to the Moon—it's a robotic scout that's helping to find the best places to explore. Future images will be even better. And by the way, those images are publicly available, so you can show them to anyone who somehow continues to believe we faked it all.

NASA operates on our hearts, on our minds, on the educational pipeline—all for one-half of one cent on the tax dollar. It's remarkable how many people think NASA's budget is bigger than that. I want to start a movement where government agencies get paid the budget people think they're getting. NASA's budget would rise by a factor of at least ten.

Space Tweet #22
NASA costs Americans half a penny on a tax dollar. That fraction of a bill is not wide enough from the edge to reach the ink
Jul 8, 2011 11:05 AM

That people think NASA's budget is huge is a measure of the visibility of every NASA dollar that gets spent. An extraordinary compliment that I wouldn't give up for anything, lest we stop advancing in all the areas Americans have come to value in the twentieth and twenty-first centuries.

For me, an interesting feature about NASA is its ten centers scattered across the country. If you grow up near one of them, you have either a relative or a friend who works for NASA. Working for NASA is a point of pride in those communities, and that sense of participation, of common journey, is something that makes this agency an enterprise for the entire nation, not simply for the select few.

Some engineers and administrators and other workers from the Apollo era still work at NASA today—though likely not for much longer. We are destined to lose them. Many, many people besides the astronauts contributed in essential ways to the Apollo era. Think of it as a pyramid. At the base are thousands of engineers and scientists, laying the groundwork for the Moon voyages. As you work your way up the pyramid, the astronauts are at the top—the brave ones putting their lives at risk. But in doing so, they place their trust in what the rest of that pyramid provides. And what sustains the base of that pyramid, keeping it broad and sturdy, is inspiration of the coming generation.

••• CHAPTER TWENTY-ONE

HOW TO REACH THE SKY*

In daily life you rarely need to think about propulsion, at least the kind that gets you off the ground and keeps you aloft. You can get around just fine without booster rockets simply by walking, running, roller-blading, taking a bus, or driving a car. All those activities depend on friction between you (or your vehicle) and Earth's surface.

When you walk or run, friction between your feet and the ground enables you to push forward. When you drive, friction between the rubber wheels and the pavement enables the car to move forward. But try to run or drive on slick ice, where there's hardly any friction, and you'll slip and slide and generally embarrass yourself as you go nowhere fast.

For motion that doesn't engage Earth's surface, you'll need a vehicle equipped with an engine stoked with massive quantities of fuel. Within the atmosphere, you could use a propeller-driven engine or a jet, both fed by fuel that burns the free supply of oxygen provided by the air. But if you're hankering to cross the airless vacuum of space, leave the props and jets at home and look for a propulsion mechanism that requires no friction and no chemical help from the air.

One way to get a vehicle to leave our planet is to point its nose upward, aim its engine nozzles downward, and swiftly sacrifice a goodly amount of the vehicle's total mass. Release that mass in one direction, and the vehicle recoils in the other. Therein lies the soul of propulsion. The mass released

---

* Adapted from "Fueling Up," *Natural History*, June 2005.

by a spacecraft is hot, spent fuel, which produces fiery, high-pressure gusts of exhaust that channel out the vehicle's hindquarters, enabling the spacecraft to ascend.

Propulsion exploits Isaac Newton's third law of motion, one of the universal laws of physics: for every action, there is an equal and opposite reaction. Hollywood, you may have noticed, rarely obeys that law. In classic Westerns, the gunslinger stands flat-footed, barely moving a muscle as he shoots his rifle. Meanwhile, the ornery outlaw that he hits sails backward off his feet, landing butt first in the feeding trough—clearly a mismatch between action and reaction. Superman exhibits the opposite effect: he doesn't recoil even slightly as bullets bounce off his chest. Arnold Schwarzenegger's character the Terminator was truer to Newton than most: every time a shotgun blast hit the cybernetic menace, he recoiled—a bit.

Spacecraft, however, can't pick and choose their action shots. If they don't obey Newton's third law, they'll never get off the ground.

Realizable dreams of space exploration took off in the 1920s, when the American physicist and inventor Robert H. Goddard got a small liquid-fueled rocket engine off the ground for nearly three seconds. The rocket rose to an altitude of forty feet and landed 180 feet from its launch site.

But Goddard was hardly alone in his quest. Several decades earlier, around the turn of the twentieth century, a Russian physicist named Konstantin Eduardovich Tsiolkovsky, who earned his living as a provincial high school teacher, had already set forth some of the basic concepts of space travel and rocket propulsion. Tsiolkovsky conceived of, among other things, multiple rocket stages that would drop away as the fuel in them was used up, reducing the weight of the remaining load and thus maximizing the capacity of the remaining fuel to accelerate the craft. He also came up with the so-called rocket equation, which tells you just how much fuel you'll need for your journey through space.

Nearly half a century after Tsiolkovky's investigations came the forerunner of modern spacecraft, Nazi Germany's V-2 rocket. The V-2 was conceived and designed for war, and was first used in combat in 1944, principally to terrorize London. It was the first rocket to target cities that lay beyond its own horizon. Capable of reaching a top speed of about 3,500 miles an hour, the V-2 could go a few hundred miles before plummeting back to Earth's surface in a deadly free fall from the edge of space.

To achieve a full orbit of Earth, however, a spacecraft must travel five times faster than the V-2, a feat that, for a rocket of the same mass as the V-2, requires no less than twenty-five times the V-2's energy. And to escape from Earth orbit altogether and head out toward the Moon, Mars, or beyond, the craft must reach 25,000 miles an hour. That's what the Apollo missions did in the 1960s and 1970s to get to the Moon—a trip requiring at least another factor of two in energy.

And that represents a phenomenal amount of fuel.

Because of Tsiolkovsky's unforgiving rocket equation, the biggest problem facing any craft heading into space is the need to boost "excess" mass in the form of fuel, most of which is the fuel required to transport the fuel it will burn later in the journey. And the spacecraft's weight problems grow exponentially. The multistage vehicle was invented to soften this problem. In such a vehicle, a relatively small payload—such as the Apollo spacecraft, an Explorer satellite, or the space shuttle—gets launched by huge, powerful rockets that drop away sequentially or in sections when their fuel supplies become exhausted. Why tow an empty fuel tank when you can just dump it and possibly reuse it on another flight?

Take the Saturn V, a three-stage rocket that launched the Apollo astronauts toward the Moon. It could almost be described as a giant fuel tank. The Saturn V and its human cargo stood thirty-six stories tall, yet the three astronauts returned to Earth in an itty-bitty, one-story capsule. The first stage dropped away two and a half minutes after liftoff, once the vehicle had been boosted off the ground and was moving at about 9,000 feet per second (about 6,000 miles per hour). Stage two dropped away about six minutes later, once the vehicle was moving at about 23,000 feet per second (almost 16,000 miles per hour). Stage three had a more complicated life, with several episodes of fuel burning: the first to accelerate the vehicle into Earth orbit, the next to get out of Earth orbit and head toward the Moon, and, three days later, one or two more thrusts to slow down and pull into lunar orbit. At each stage, the craft got progressively smaller and lighter, which means that the remaining fuel could do more with less.

From 1981 to 2011, NASA used the space shuttle for missions a few hundred miles above our planet: low Earth orbit. The shuttle has three main parts: a stubby, airplanelike "orbiter" that holds the crew, the payload, and the three main engines; an immense external fuel tank that holds more than half a million gallons of self-combustible liquid; and two "solid rocket boosters," whose two million pounds of rubbery aluminum-based fuel generate 85 percent of the thrust needed to get the giant off the ground. On the launchpad the shuttle weighs four and a half million pounds. Two minutes after launch, the boosters have finished their work and drop away into the ocean, to be fished out of the water and reused. Six minutes later, just before the shuttle reaches orbital speed, the now-empty external tank drops off and disintegrates as it reenters Earth's atmosphere. By the time the shuttle reaches orbit, 90 percent of its launch mass has been left behind.

Space Tweet #23
Main shuttle tank in use until orbit – long after atmospheric O2 is available to burn. So must carry its own O2.
May 14, 2010 3:03 AM

Now that you're launched, how about slowing down, landing gently, and one day returning home? Fact is, in empty space, slowing down takes as much fuel as speeding up.

Familiar, earthbound ways to slow down require friction. On a bicycle, the rubber pincers on the hand brake squeeze the wheel rim; on a car, the brake pads squeeze against the wheels' rotors, slowing the rotation of the four rubber tires. In those cases, stopping requires no fuel. To slow down and stop in space, however, you must turn your rocket nozzles backward, so that they point in the direction of motion, and ignite the fuel you've dragged all that distance. Then you sit back and watch your speed drop as your vehicle recoils in reverse.

To return to Earth after your cosmic excursion, rather than using fuel to slow down, you could do what the space shuttle does: glide back to Earth unpowered, and exploit the fact that our planet has an atmosphere, a source of friction. Instead of using all that fuel to slow down the craft before reentry, you could let the atmosphere slow it down for you.

Space Tweets #24–#27

Discovery Orbiter re-enters today. From 17,000mph to 0mph in an hour. Relies on air resistance (aerobraking) to slow down

Mar 9, 2011 8:30 AM

Will take 3/4 of a trip around Earth for atmosphere to drop Discovery out of the sky & land safely as a glider at Kennedy, FL

Mar 9, 2011 10:54 AM

After the Shuttle drops below sound speed (Mach 1) it's just a fat, stubby glider coming in for a landing

Mar 9, 2011 11:51 AM

Welcome home Discovery. 39 missions, 365 days & 148,221,675 miles on the odometer

Mar 9, 2011 11:59 AM

One complication, though, is that the craft is traveling much faster during its home stretch than it was during its launch. It's dropping out of a seventeen-thousand-mile-an-hour orbit and plunging toward Earth's surface, so heat and friction are much bigger problems at the end of the journey than at the beginning. One solution is to sheathe the leading surface of the craft in a heat shield, which deals with the swiftly accumulating heat through ablation or dissipation. In ablation, the preferred method for the cone-shaped Apollo-era capsules, the heat gets carried away by shock waves in the air and a continuously peeling supply of vaporized material on the capsule's bottom. For the space shuttle and its famous tiles, dissipation is the method of choice.

Unfortunately, as we all now know, heat shields are hardly invulnerable. The seven astronauts of the Columbia space shuttle were cremated in midair on the morning of February 1, 2003, as their orbiter tumbled out of control and broke apart during reentry. They met their deaths because a chunk of foam insulation had come loose from the shuttle's huge fuel tank during the launch and had pierced a hole in the leading shield that covered the left wing. That hole exposed the orbiter's aluminum dermis, causing it to warp and melt in the rush of superheated air.

• • •

Here's a safer idea for the return trip: Why not put a filling station in Earth orbit? When it's time for the shuttle to come home, you attach a new set of tanks and fire them at full throttle, backward. The shuttle slows to a crawl, drops into Earth's atmosphere, and just flies home like an airplane. No friction. No shock waves. No heat shields.

But how much fuel would that take? Exactly as much fuel as it took to get the thing up there to begin with. And how might all that fuel reach the orbiting filling station that could service the shuttle's needs? Presumably it would be launched there, atop some other skyscraper-high rocket.

Think about it. If you wanted to drive from New York to California and back again, and there were no gas stations along the way, you'd have to tug a truck-size fuel tank. But then you'd need an engine strong enough to pull a truck, so you'd need to buy a much bigger engine. Then you'd need even more fuel to drive the car. Tsiolkovsky's rocket equation eats your lunch every time.

In any case, slowing down or landing isn't only about returning to Earth. It's also about exploration. Instead of just passing the far-flung planets in fleeting "flybys," a mode that characterized an entire generation of NASA space probes, the craft ought to spend some time getting to know those distant worlds. But it takes extra fuel to slow down and pull into orbit. Voyager 2, for instance—launched in August 1977—has spent its entire life coasting. After gravity assists, first from Jupiter and then from Saturn (the gravity assist is the poor man's propulsion mechanism), Voyager 2 flew past Uranus in January 1986 and past Neptune in August 1989. For a spacecraft to spend a dozen years reaching a planet and then spend only a few hours there collecting data is like waiting two days in line to see a rock concert that lasts six seconds. Flybys are better than nothing, but they fall far short of what a scientist really wants to do.

On Earth, a fill-up at the local gas station has become a pricey activity. Plenty of smart scientists have spent plenty of years inventing and developing alternative fuels that might one day see widespread use. And plenty of other smart scientists are doing the same for propulsion.

The most common forms of fuel for spacecraft are chemical substances: ethanol, hydrogen, oxygen, monomethyl hydrazine, powdered aluminum. But unlike airplanes, which burn fuel by drawing oxygen through their engines, spacecraft have no such luxury; they must bring the whole chemical equation along with them. So they carry not only the fuel but an oxidizer as well, kept separate until valves bring them together. The ignited, high-temperature mixture then creates high-pressure exhaust, all in the service of Newton's third law of motion.

Bummer. Even ignoring the free "lift" a plane gets from air rushing over its specially shaped wings, pound for pound any craft whose agenda is to leave the atmosphere must carry a much heavier fuel load than an airplane does. The V-2's fuel was ethanol and water; the Saturn V's fuel was kerosene for the first stage and liquid hydrogen for the second stage. Both rockets used liquid oxygen as the oxidizer. The space shuttle's main engine, which had to work above the atmosphere, used liquid hydrogen and liquid oxygen.

Wouldn't it be nice if the fuel itself carried more punch than it does? If you weigh 150 pounds and you want to launch yourself into space, you'll need 150 pounds of thrust under your feet (or spewed forth from a jet pack) just to weigh nothing. To actually launch yourself, anything more than 150 pounds of thrust will do, depending on your tolerance for acceleration. But wait. You'll need even more thrust than that to account for the weight of the unburned fuel you're carrying. Add more thrust than that, and you'll accelerate skyward.

Space Tweet #28

At a fine Italian restaurant this evening. Served grappa at meal's end. NASA should study it as a replacement rocket fuel

Dec 7, 2010 12:27 AM

The space mavens' perennial goal is to find a fuel source that packs astronomical levels of energy into the smallest possible volumes. Because chemical fuels use chemical energy, there's a limit to how much thrust they can provide, and that limit comes from the stored binding energies within molecules. Even given those limitations, there are several innovative options. After a vehicle rises beyond Earth's atmosphere, propulsion need not come from burning vast quantities of chemical fuel. In deep space, the propellant can be small amounts of ionized xenon gas, accelerated to enormous speeds within a new kind of engine. A vehicle equipped with a reflective sail can be pushed along by the gentle pressure of the Sun's rays, or even by a laser stationed on Earth or on an orbiting platform. And within a decade or so, a perfected, safe nuclear reactor will make nuclear propulsion possible—the rocket designer's dream engine. The energy it generates will be orders of magnitude more than chemical fuels can produce.

While we're getting carried away with making the impossible possible, what we really want is the antimatter rocket. Better yet, we'd like to arrive at a new understanding of the universe, to enable journeys that exploit wormhole shortcuts in the fabric of space and time. When that happens, the sky will no longer be the limit.

••• CHAPTER TWENTY-TWO

THE LAST DAYS OF THE SPACE SHUTTLE

May 16, 2011: The Final Launch of Endeavour

Space Tweets #29–#36

8:29 AM

If camera-coverage enables, six cool things to look for just seconds before ignition of the SolidRocketBoosters...

8:30 AM

1) Orbiter's steering flaps jiggle back and forth – a final reminder that they can angle the way they're supposed to

8:32 AM

2) The Orbiter's 3 rocket nozzles gimbal to & fro – a final reminder that they can aim the way they're supposed to

8:33 AM

3) Sparks spray onto launch pad – they burn away any potentially flammable hydrogen gathered there from the main engine

8:35 AM

4) Water Tower dumps a swimming-pool's worth onto the launch pad – H2O absorbs sound vibrations, preventing damage to craft

8:37 AM

5) "Main Engine Start" – Orbiter's 3 nozzles ignite, take aim, and force shuttle to tip forward. Bolts still hold her down

8:38 AM

6) "3 - 2 - 1 – Liftoff" – SolidRocketBoosters ignite, tipping Shuttle straight upwards again. Bolts explode. Craft ascends

9:18 AM

In case you wondered: Space Shuttle Endeavour gets a British spelling because it's named for Captain Cook's ship

June 1, 2011: The Final Return of Endeavour

Space Tweets #37–#45

1:20 AM

Just an FYI: To land, space shuttle Endeavour must lose all the energy of motion that it gained during launch

1:30 AM

Shuttle now executing a "de-orbit burn" dropping its path low enough to meet scads of motion-impeding air molecules

2:00 AM

As Endeavour dips into Earth's atmosphere, the surrounding air heats up, whisking away the Shuttle's energy of motion

2:10 AM

As Endeavour's speed slows, it drops lower in Earth's atmosphere, encountering an ever-increasing density of air molecules

2:20 AM

Protective Shuttle tiles reach thousands of degrees (F), persistently radiating heat away. Shielding the astronauts within

2:30 AM

For most of Endeavour's re-entry, it's a ballistic brick falling from the sky. Below the speed of sound, it's aerodynamic

2:34 AM

Kennedy Space Center's Shuttle's landing strip is 15,000 feet long. Long enough for the brakeless Orbiter to coast to a stop

2:35 AM

Welcome home Astronauts: 248 orbits, 6,510,221 miles. Well done Endeavour: 25 missions. 4671 orbits, 123,883,151 miles

9:10 AM

Einstein's relativity shows that Endeavour astronauts moved 1/2000 sec into the future during their stay in orbit

July 8–21, 2011: The Final Journey of Atlantis & the End of the Shuttle Era

Space Tweets #46–#51

Jul 8 9:54 am

Shuttle mission in the film "Space Cowboys" was STS-200. With the launch of Atlantis, the actual program reaches only STS-135

Jul 8 10:25 am

Space Arithmetic: Mercury + Gemini + Apollo = 10 years. Shuttle = 30 years

Jul 8 10:52 am

Just an FYI: Human access to space doesn't end with the Shuttle era, only American access. China and Russia still go there

Jul 8 11:24 am

Apollo in 1969. Shuttle in 1981. Nothing in 2011. Our space program would look awesome to anyone living backwards thru time

Jul 21 5:42 am

Worried about privatization of access to Earth orbit? Overdue by decades. NASA needs to look beyond, where it belongs.

Jul 21 5:49 am

Lament not the shuttle's end, but the absence of rockets to supplant it. Who shed a tear when Gemini ended? Apollo awaited us

••• CHAPTER TWENTY-THREE

PROPULSION FOR DEEP SPACE[*]

Launching a spacecraft is now a routine feat of engineering. Attach the fuel tanks and rocket boosters, ignite the chemical fuels, and away it goes.

But today's spacecraft quickly runs out of fuel. So, left to itself, it cannot slow down, stop, speed up, or make serious changes in direction. With its trajectory choreographed entirely by the gravity fields of the Sun, the planets, and their moons, the craft can only fly past its destination, like a fast-moving tour bus with no stops on its itinerary—and the riders can only glance at the passing scenery.

If a spacecraft can't slow down, it can't land anywhere without crashing, which is not a common objective of aerospace engineers. Lately, however, engineers have been getting clever about fuel-deprived craft. In the case of the Mars rovers, their stupendous speed toward the Red Planet was slowed by aerobraking through the Martian atmosphere. That meant they could land with the help of nothing more than heat shields, parachutes, and airbags.

Today, the biggest challenge in aeronautics is to find a lightweight and efficient means of propulsion, whose punch per pound greatly exceeds that of conventional chemical fuels. With that challenge met, a spacecraft could leave the launchpad with fuel reserves onboard, and scientists could think more about celestial objects as places to visit than as planetary peep shows.

---

* Adapted from "Heading Out," *Natural History*, July–August 2005.

Fortunately, human ingenuity doesn't often take no for an answer. Legions of engineers are ready to propel us and our robotic surrogates into deep space with a variety of innovative engines. The most efficient among them would use antimatter as fuel. When you bring matter and antimatter into contact with each other, you convert all their mass into propulsion energy, just as *Star Trek*'s antimatter engines did. Some physicists even dream of traveling faster than the speed of light by somehow tunneling through warps in the fabric of space and time. *Star Trek* didn't miss that one either: the warp drives on the starship USS Enterprise were what enabled Captain Kirk and his crew to speed across the galaxy during the TV commercials.

Acceleration can be gradual and prolonged, or it can come from a brief, spectacular blast. Only a major blast can propel a spacecraft off the ground. You've got to have at least as many pounds of thrust as the weight of the craft itself. Otherwise, the thing will just sit there on the pad. After that, if you're not in a big rush—and if you're sending cargo rather than crew to the distant reaches of the solar system—there's no need for spectacular acceleration.

In October 1998 an eight-foot-tall, half-ton spacecraft called Deep Space 1 launched from Cape Canaveral, Florida. During its three-year mission, Deep Space 1 tested a dozen innovative technologies, including a propulsion system equipped with ion thrusters—the kind of system that becomes useful at great distances from the launchpad, where low but sustained acceleration eventually yields very high speeds.

Ion-thruster engines do what conventional spacecraft engines do: they accelerate propellant (in this case, a gas) to very high speeds and channel it out a nozzle. In response, the engine, and thus the rest of the spacecraft, recoils in the opposite direction. You can do this science experiment yourself: While you're standing on a skateboard, let loose a CO_2 fire extinguisher (purchased, of course, for this purpose). The gas will go one way; you and the skateboard will go the other way.

But ion thrusters and ordinary rocket engines part ways in their choice of propellant and their source of the energy that accelerates it. Deep Space 1 used electrically charged (ionized) xenon gas as its propellant, rather than the liquid hydrogen-oxygen combo burned in the space shuttle's main engine. Ionized gas is easier to manage than explosively flammable chemicals. Plus, xenon happens to be a noble gas, which means it won't corrode or otherwise interact chemically with anything. For sixteen thousand hours, using less than four ounces of propellant a day, Deep Space 1's foot-wide, drum-shaped engine accelerated xenon ions across an electric field to speeds of twenty-five miles per second and spewed them from its nozzle. As anticipated, the recoil per pound of fuel was ten times greater than that of conventional rocket engines.

In space as on Earth, however, there is no such thing as a free lunch—not to mention a free launch. Something had to power those ion thrusters on Deep Space 1. Some investment of energy had to first ionize the xenon atoms and then accelerate them. That energy came from electricity, courtesy of the Sun.

For touring the inner solar system, where light from the Sun is strong, the spacecraft of tomorrow can use solar panels—not for the propulsion itself, but for the electric power needed to drive the equipment that manages the propulsion. Deep Space 1, for instance, had folding solar "wings" that, when fully extended, spanned almost forty feet—about five times the height of the spacecraft itself. The arrays on them were a combination of 3,600 solar cells and more than seven hundred cylindrical lenses that focused sunlight on the cells. At peak power, their collective output was more than two thousand watts, enough to operate only a hair dryer or two on Earth but plenty for powering the spacecraft's ion thrusters.

Other, more familiar spacecraft—such as the deorbited and disintegrated Soviet space station Mir and the sprawling International Space Station (ISS)—have also depended on the Sun for the power to operate their electronics. Orbiting about 250 miles above Earth, the ISS carries more than an acre's worth of solar panels. For about a third of every ninety-minute orbit, as Earth eclipses the Sun, the station orbits in darkness. So by day, some of the collected solar energy gets channeled into storage batteries for later use during dark hours.

Although neither Deep Space 1 nor the ISS has used the Sun's rays to propel itself, direct solar propulsion is far from impossible. Consider the solar sail, a gossamer, somewhat kitelike form of space propulsion that, once aloft, will accelerate because of the collective thrust of the Sun's photons, or particles of light, continually reflecting off the sail's shiny surfaces. As they bounce, the photons induce the craft to recoil. No fuel. No fuel tanks. No exhaust. No mess. You can't get greener than that.

Having envisioned the geosynchronous satellite, Sir Arthur C. Clarke went on to envision the solar sail. For his 1964 story "The Wind from the Sun," he created a character who described how it would work:

> Hold your hands out to the sun. What do you feel? Heat, of course. But there's pressure as well—though you've never noticed it, because it's so tiny. Over the area of your hands, it only comes to about a millionth of an ounce. But out in space, even a pressure as small as that can be important—for it's acting all the time, hour after hour, day after day. Unlike rocket fuel, it's free and unlimited. If we want to, we can use it; we can build sails to catch the radiation blowing from the sun.

In the 1990s, a group of US and Russian rocket scientists who preferred to collaborate rather than contribute to mutual assured destruction (aptly known as MAD) began working on solar sails through a privately funded collaboration led by the Planetary Society. The fruit of their labor, Cosmos 1, was an engineless, 220-pound spacecraft shaped like a supersize daisy. This celestial sailboat folded inside an unarmed intercontinental ballistic missile left over from the Soviet Union's Cold War arsenal and was launched from a Russian submarine. Cosmos 1 had a computer at its center and eight reflective, triangular sail blades made of 0.0002-inch-thick Mylar—much thinner than a cheap trash bag—and reinforced with aluminum. When unfurled in space, each blade would extend fifty feet and could be individually angled to steer and sail the craft. Alas, the rocket engine failed little more than a minute after launch, and the furled sail itself, apparently still attached to the rocket, fell into the Barents Sea.

But engineers don't stop working just because their early efforts fail. Today not only the Planetary Society but also NASA, the US Air Force, the European Space Agency, universities, corporations, and start-ups are enthusiastically investigating designs and uses of solar sails. Philanthropists have come forth with million-dollar donations. International conferences on solar sailing now take place. And in 2010, space sailors celebrated their community's first true success: a 650-square-foot, 0.0003-inch-thick sail named IKAROS (Interplanetary Kite-craft Accelerated by Radiation Of the Sun), designed and operated by the Japan Aerospace Exploration Agency, JAXA. The sail entered solar orbit on May 21, finished unfurling itself on June 11, and passed Venus on December 8. Meanwhile, the Planetary Society anticipates a launch of its LightSail-1, and NASA is working on a miniature demonstration craft named Nano-Sail-D, which may point the way toward using solar sails as parachutes to tow defunct satellites out of orbit and out of harm's way.

So let's look on the sunny side. Having entered space, a lightweight solar sail could, after a couple of years, accelerate to a hundred thousand miles an hour. That's the remarkable effect of a low but steady acceleration. Such a craft could escape from Earth orbit (where it was lofted by conventional rockets) not by aiming for a destination but by cleverly angling its blades, as does a sailor on a ship, so that it ascends to ever larger orbits around Earth. Eventually its orbit could become the same as that of the Moon, or Mars, or something beyond.

Obviously a solar sail would not be the transportation of choice for anybody in a hurry to receive supplies, but it would certainly be fuel efficient. If you wanted to use it as, say, a low-cost food-delivery van, you could load it up with dried fruit, ready-to-eat breakfast cereals, Twinkies, Cool Whip, and other edible items of extremely high shelf life. And as the craft sailed into sectors where the Sun's light is feeble, you could help it along with a laser, beamed from Earth, or with a network of lasers stationed across the solar system.

Speaking of regions where the Sun is dim, suppose you wanted to park a space station in the outer solar system—at Jupiter, for instance, where sunlight is only 1/27 as intense as it is here on Earth. If your Jovian space station required the same amount of solar power as the completed International Space Station, your panels would have to cover twenty-seven acres. So you would now be laying solar arrays over an area bigger than twenty football fields. I think not. To do complex science in deep space, to enable explorers (or settlers) to spend time there, to operate equipment on the surfaces of distant planets, you must draw energy from sources other than the Sun.

• • •

Since the early 1960s, space vehicles have commonly relied on the heat from radioactive plutonium as an electrical power supply. Several of the Apollo missions to the Moon, as well as Pioneer 10 and 11 (now about ten billion miles from Earth and destined for interstellar space), Viking 1 and 2 (to Mars), Voyager 1 and 2 (also destined for interstellar space and, in the case of Voyager 1, farther along than the Pioneers), Ulysses (to the Sun), Cassini (to Saturn), and New Horizons (to Pluto and the Kuiper Belt), among others, have all used plutonium for their radioisotope thermoelectric generators, or RTGs. An RTG is a long-lasting source of nuclear power. Much more efficient, and much more energetic, would be a nuclear reactor that could supply both power and propulsion.

Nuclear power in any form, of course, is anathema to some people. Good reasons for this view are not hard to find. Inadequately shielded plutonium and other radioactive elements pose great danger; uncontrolled nuclear chain reactions pose even greater danger. And it's easy to draw up a list of proven and potential disasters: the radioactive debris spread across northern Canada in 1978 by the crash of the nuclear-powered Soviet satellite Cosmos 954; the partial meltdown in 1979 at the Three Mile Island nuclear power plant on the Susquehanna River near Harrisburg, Pennsylvania; the explosion at the Chernobyl nuclear power plant in 1986 in what is now Ukraine; the plutonium in old RTGs currently lying in (and occasionally stolen from) remote, decrepit lighthouses in northwestern Russia. The failure of the Fukushima Daiichi nuclear power plant on Japan's northeast coast, struck by a 9.0 earthquake and then inundated by a horrific tsunami in March 2011, renewed every fear. Citizens' organizations such as the Global Network Against Weapons and Nuclear Power in Space remember these and other similar events.

But so do the scientists and engineers who worked on NASA's Project Prometheus.

Rather than deny the risks of nuclear devices, NASA turned its attention to maximizing safeguards. In 2003 the agency charged Project Prometheus with developing a small nuclear reactor that could be safely launched and could power long and ambitious missions to the outer solar system. Such a reactor was to provide onboard power and could drive an electric engine with ion thrusters—the same kind of propulsion tested in Deep Space 1.

To appreciate the advance of technology, consider the power output of the RTGs that drove the experiments on the Vikings and Voyagers. They supplied less than a hundred watts, about what your desk lamp uses. The RTGs on Cassini do a bit better, nearly three hundred watts: about the power required by a small kitchen appliance. The nuclear reactor that should have emerged from Prometheus was slated to yield ten thousand watts of usable power for its scientific instruments, enough to drive a rock concert.

To exploit the Promethean advance, an ambitious scientific mission was proposed: the Jupiter Icy Moons Orbiter, or JIMO. Its destinations were Callisto, Ganymede, and Europa—three of the four moons of Jupiter discovered by Galileo in 1610. (The fourth, Io, is studded with active, flaming hot volcanoes.) The lure of the three frigid Galilean moons was that beneath their thick crust of ice might lie vast reservoirs of liquid water that harbor, or once harbored, life.

Endowed with ample onboard propulsion, JIMO would do a "flyto," rather than a flyby, of Jupiter eight years after launch. It would pull into orbit and systematically visit one moon at a time, perhaps even deploying landers. Powered by ample onboard electricity, suites of scientific instruments would study the moons and send data back to Earth via high-speed broadband channels. Besides efficiency, a big attraction would be safety, both structural and operational. The spacecraft would be launched with ordinary rockets, and its nuclear reactor would be launched "cold"—not until JIMO had reached escape velocity and was well out of Earth orbit would the reactor be turned on.

Sounded good. But Prometheus/JIMO died after barely having lived, becoming what a committee constituted by the National Research Council's Space Studies Board and Aeronautics and Space Engineering Board termed, in a 2008 report titled *Launching Science,* a "cautionary tale." Formally started in March 2003 as a science program, it was transferred within the year to NASA's newly established Exploration Systems Mission Directorate. Less than a year and a half later, in the summer of 2005, after spending nearly $464 million (plus tens of millions of dollars simply to fund the preparation of the contractors' bids), NASA canceled the program. Over the succeeding months, $90 million of its $100 million budget went for closeout costs on the canceled contracts. All that money, and yet no spacecraft and no scientific findings. Prometheus/JIMO thus stands, write the authors of *Launching Science,* as "an example of the risks associated with pursuing ambitious, expensive space science missions."

Risks, cancellations, and failures are just part of the game. Engineers expect them, agencies resist them, accountants juggle them. Cosmos 1 may have dropped into the sea, and Prometheus/JIMO may have died in the cradle, but they yielded valuable technical lessons. So, hopeful cosmic travelers have no reason to stop trying or planning or dreaming about how to navigate in deep space. Today's term of art is "in-space propulsion," and plenty of people are still avidly pursuing its possibilities, including NASA. More efficient rockets are one approach, and so NASA is developing advanced high-temperature rockets. Better thrusters are another approach, and so NASA now has the NEXT (NASA's Evolutionary Xenon Thruster) Ion Propulsion System, a few steps up from the system on Deep Space 1. Then there are the aforementioned solar sails. The goals of all of these technologies, individually and/or in combination, are to cut down the travel time to distant celestial bodies, increase the potential range and weight of the scientific payload, and reduce the costs.

Someday there might be wackier ways to explore within and beyond our solar system. The folks at NASA's now-defunct Breakthrough Propulsion Physics Project, for instance, were dreaming of how to couple gravity and electromagnetism, or tap the zero-point energy states of the quantum vacuum, or harness superluminal quantum phenomena. Their inspiration came from such tales as *From the Earth to the Moon*, by Jules Verne, and the adventures of Buck Rogers, Flash Gordon, and *Star Trek*. It's okay to think about this sort of thing from time to time. But, in my opinion, though it's possible not to have read enough science fiction in one's lifetime, it's also possible to have read too much of it.

My favorite science-fiction engine is the antimatter drive. It's 100 percent efficient: put a pound of antimatter together with a pound of matter, and they turn into a puff of pure energy, with no by-products. Antimatter is real. Credit the twentieth-century British physicist Paul A. M. Dirac for conceiving of it in 1928, and the American physicist Carl D. Anderson for discovering it five years later.

The science part of antimatter is fine. It's the science-fiction part that presents a small problem. How do you store the stuff? Behind whose spaceship cabin or under whose bunk bed would the canister of antimatter be kept? And out of what substance would the canister be made? Antimatter and matter annihilate each other on contact, so keeping antimatter around requires portable matterless containers, such as magnetic fields shaped into magnetic bottles. Unlike the fringe propulsion ideas, where engineering chases the bleeding edge of physics, the antimatter problem is ordinary physics chasing the bleeding edge of engineering.

So the quest continues. Meanwhile, next time you're watching a movie in which a captured spy is being questioned, think about this: The questioners hardly ever ask about agricultural secrets or troop movements. With an eye to the future, they ask about the secret rocket formula, the transportation ticket to the final frontier.

••• CHAPTER TWENTY-FOUR

BALANCING ACTS*

The first manned spacecraft ever to leave Earth orbit was Apollo 8. This achievement remains one of the most unappreciated firsts of the twentieth century. When that moment arrived, the astronauts fired the third and final stage of their mighty Saturn V rocket, and the spacecraft and its three occupants rapidly reached a speed of nearly seven miles per second. As the laws of physics show, just by reaching Earth orbit the astronauts had already acquired half the energy needed to reach the Moon.

After Apollo 8's third stage fired, engines were no longer necessary except to tune the midcourse trajectory so that the astronauts did not miss the Moon entirely. For most of its nearly quarter-million-mile journey from Earth to the Moon, the spacecraft gradually slowed as Earth's gravity continued to out-tug the Moon's gravity. Meanwhile, as the astronauts neared the Moon, *its* force of gravity grew stronger and stronger. Obviously there had to be a spot en route where the Moon's and Earth's opposing forces of gravity balanced precisely. And when the command module drifted across that point in space, its speed increased once again, and it accelerated toward the Moon.

If gravity were the only force to be reckoned with, then that spot would be the only place in the Earth–Moon system where the opposing forces

* Adapted from "The Five Points of Lagrange," *Natural History*, April 2002.

cancel. But Earth and the Moon revolve around a common center of gravity, which lives about a thousand miles beneath Earth's surface along the length of an imaginary line connecting the center of Earth to the center of the Moon.

When moving objects are pulled in circles of any size and at any speed, they create a new sensation that pushes outward, away from the center of rotation. Your body feels this "centrifugal" force when you make a sharp turn in your car or when you survive amusement-park attractions that turn in circles. In a classic example of these nausea-inducing rides, you stand along the edge of a large circular platter, with your back against a perimeter wall. As the ride spins, rotating faster and faster, you feel a stronger and stronger force pinning you against the wall. It's the sturdy wall that prevents you from being flung through the air. Soon you can't move. That's when they drop the floor from below your feet and turn the thing sideways and upside down. When I rode one of these as a kid, the force was so great that I could barely move my fingers: they stuck to the wall along with the rest of me. (If you actually got sick on such a ride and you turned your head sideways, the vomit would fly off at a tangent. Or it might get stuck to the wall. Worse yet, if you didn't turn your head, it might not make it out of your mouth, owing to the extreme centrifugal forces acting in the opposite direction. Come to think of it, I haven't seen this particular ride anywhere lately.)

Centrifugal forces arise as the simple consequence of an object's tendency to travel in a straight line after being set in motion, and so are not true forces at all. But you can use them in calculations as though they were. The brilliant eighteenth-century French mathematician Joseph-Louis Lagrange discovered spots in the rotating Earth–Moon system where the gravity of Earth, the gravity of the Moon, and the centrifugal forces of the rotating system all balance. These special locations are known as the points of Lagrange, and there are five of them.

The first point of Lagrange (sensibly called L1) falls slightly closer to Earth than the point of pure gravitational balance. Any object placed at L1 can orbit the Earth–Moon center of gravity with the same monthly period as the Moon's orbit and will appear to be locked in place along the Earth–Moon line. Although all forces cancel there, L1 is a point of precarious equilibrium. If the object drifts *away from* the Earth–Moon line in any direction, the combined effect of the three forces will return it to its former position. But if the object drifts *along* the Earth–Moon line ever so slightly, it will irreversibly fall toward either Earth or the Moon. It's like a cart atop a mountain, barely balanced, a hair's width away from rolling down one side or the other.

The second and third Lagrangian points (L2 and L3) also lie on the Earth–Moon line, but L2 lies beyond the Moon, while L3 lies far beyond Earth in the opposite direction. Once again, the three forces—Earth's gravity, the Moon's gravity, and the centrifugal force of the rotating system—cancel in concert. And once again, an object placed in either spot can orbit the Earth–Moon center of gravity in a lunar month. The gravitational balance points at L2 and L3 are quite broad. So if you find yourself drifting down to Earth or the Moon, a tiny investment in fuel will bring you right back to where you were.

Although L1, L2, and L3 are respectable space places, the award for best Lagrangian points must go to L4 and L5. One of them lives far off to one side of the Earth–Moon centerline, while the other lives far off to the opposite side, and each of them represents one vertex of an equilateral triangle, with Earth and the Moon serving as the other two vertices. At L4 and L5, as with their first three siblings, forces are in equilibrium. But unlike the first three Lagrangian points, which enjoy only unstable equilibrium, the equilibria at L4 and L5 are stable. No matter which direction you lean, no matter which direction you drift, the forces prevent you from leaning farther, as though you were at the bottom of a bowl-shaped crater surrounded by a high, sloped rim. So, for both L4 and L5, if an object is not located exactly where all forces cancel, then its position will oscillate around the point of balance, in paths called librations. (Not to be confused with the particular spots on Earth's surface where one's mind oscillates from ingested libations.) These librations are equivalent to the back-and-forth path a ball takes when it rolls down one hill yet doesn't pick up enough speed to climb the next.

More than just orbital curiosities, L4 and L5 represent special areas where one might decide to establish space colonies. All you need do is ship some raw construction materials to the area (having mined them not only from Earth but perhaps from the Moon or an asteroid); leave them in place, since there's no risk of their drifting away; and return later with more supplies. Once you've collected all your materials in this zero-G environment, you could build an enormous space station—tens of miles across—with very little stress on the materials themselves. By rotating the station, you would induce centrifugal forces that simulate Earth gravity for its hundreds (or thousands) of residents and their farm animals.

In 1975, Keith and Carolyn Henson founded the L5 Society to carry out exactly those plans, although the society is best remembered for its informal association with Princeton physics professor Gerard K. O'Neill, who promoted space habitation through such visionary writings as his 1976 book *The High Frontier: Human Colonies in Space*. The group had a single goal: "to disband the Society in a mass meeting at L5." Presumably this would be done inside the completed space habitat, during the party celebrating their mission accomplished. In 1987 the L5 Society merged with the National Space Institute to become the National Space Society, which continues today.

The idea of locating a large structure at libration points appeared as early as the early 1940s, in a series of sci-fi short stories by George O. Smith collected under the title *Venus Equilateral*. In them the author imagines a relay station at the L4 point of the Venus–Sun system. In 1961 Arthur C. Clarke would reference Lagrangian points in his novel *A Fall of Moondust*. Clarke, of course, was no stranger to special orbits. In 1945 he became the first to calculate, in a four-page memorandum, the altitude above Earth's surface at which a satellite's orbital period would exactly match the twenty-four-hour rotation period of Earth. Because a satellite with that orbit "hovers" over Earth's surface, it can serve as an ideal relay station for radio communications from one part of Earth to another. Today, hundreds of communication satellites do just that, at about 22,000 miles above Earth's surface.

As George O. Smith knew, there is nothing unique about the balance points in the rotating Earth–Moon system. Another set of five Lagrangian points exists for the rotating Sun–Earth system, as well as for any pair of orbiting bodies anywhere in the universe. For objects in low orbits, such as the Hubble, Earth continuously blocks a significant chunk of its night-sky view. However, a million miles from Earth, in the direction opposite that of the Sun, a telescope at the Sun–Earth L2 will have a twenty-four-hour view of the night sky, because it would see Earth at about the size we see the Moon in Earth's sky.

The Wilkinson Microwave Anisotropy Probe (WMAP for short), which was launched in 2001, reached the Sun–Earth L2 in a couple of months and is still librating there, having busily taken data on the cosmic microwave background—the omnipresent signature of the Big Bang. And having set aside a mere 10 percent of its total fuel, the WMAP satellite nevertheless has enough fuel to hang around this point of unstable equilibrium for nearly a century, long beyond its useful life as a data-taking space probe. NASA's next-generation space telescope, the James Webb Space Telescope (successor to the Hubble), is also being designed for the Sun–Earth L2 point. And there's plenty of room for yet more satellites to come and librate, since the real estate of the Sun–Earth L2 occupies quadrillions of cubic miles.

Another Lagrangian-loving NASA satellite, known as Genesis, librated around the Sun–Earth L1 point. This L1 lies a million miles out between Earth and the Sun. For two and a half years, Genesis faced the Sun and collected pristine solar matter, including atomic and molecular particles from the solar wind—revealing something of the contents of the original solar nebula from which the Sun and planets formed.

Given that L4 and L5 are stable points of equilibrium, one might suppose that space junk would accumulate near them, making it quite hazardous to conduct business there. Lagrange, in fact, had predicted in 1772 that space debris would be found at L4 and L5 of gravitationally powerful Sun–planet systems, and in 1906, the first member of Jupiter's double family of Trojan asteroids was discovered. We now know that gathered at the L4 and L5 points of the Sun–Jupiter system are thousands of asteroids that follow and lead Jupiter around the Sun, with periods that equal one Jovian year. As though gripped by tractor beams, these asteroids are forever held in place by the gravitational and centrifugal forces of the Sun–Jupiter system. (These asteroids, being stuck in the outer solar system and out of harm's way, pose no risk to life on Earth or to themselves.) Of course, we would expect space junk to accumulate at L4 and L5 of the Sun–Earth and Earth–Moon systems too. And it does.

• • •

As an important side benefit, interplanetary trajectories that begin at Lagrangian points require very little fuel to reach other Lagrangian points or even other planets. Unlike a launch from a planet's surface, where most of your fuel goes to lift you off the ground, a Lagrangian launch would be a low-energy affair and would resemble a ship leaving dry dock, cast into the sea with a minimal investment of fuel. Today, instead of thinking about establishing self-sustaining Lagrangian colonies of people and cows, we can think of Lagrangian points as gateways to the rest of the solar system. From the Sun–Earth Lagrangian points, you are halfway to Mars—not in distance or in time but in the all-important category of fuel consumption.

In one version of our spacefaring future, imagine filling stations at every Lagrangian point in the solar system, where travelers refill their rocket gas tanks en route to visit friends and relatives living on other planets or moons. This mode of travel, however futuristic it sounds, is not without precedent. Were it not for the gas stations scattered liberally across the United States, your automobile would require a colossal tank to drive coast to coast: most of the vehicle's size and mass would be fuel, guzzled primarily to transport the yet-to-be-consumed fuel for your cross-country trip. We don't travel that way on Earth. Perhaps the time will come when we no longer travel that way through space.

••• CHAPTER TWENTY-FIVE

HAPPY ANNIVERSARY, *STAR TREK*[*]

In 2011 *Star Trek* turned forty-five. Meanwhile, the television signals from all its broadcast episodes continue to penetrate our Milky Way galaxy at the speed of light. By now the first episode of the first season, which aired for the first time on September 8, 1966, has reached forty-five light-years from Earth, having swept past more than six hundred star systems, including Alpha Centauri, Sirius, Vega, and an ever-growing number of lesser-known stars around which we have confirmed the existence of planets.

It must have been a long wait for eavesdropping aliens. Their first encounter with Earth culture included the earliest episodes of the *Howdy Doody Show* and Jackie Gleason's *Honeymooners.* With the arrival of *Star Trek* some fifteen years later, we finally offered extraterrestrial anthropologists something in our TV waves that our species could be proud of.

In its many incarnations for television, film, and books, *Star Trek* became the most popular science-fiction series ever. Yet if you watch some of the original episodes, it's not hard to see how the show got canceled after three seasons. In any case, were it not for a million-plus letters written to NBC, the show would have been canceled after two. The *Star Trek* seasons happened to coincide with the most triumphant years (1966–69) of the space program as well as America's bloodiest years of the Vietnam War and

---

* Adapted from "The Science of Trek," in Stephen Reddicliffe, ed., *TV Guide—Star Trek 35th Anniversary Tribute: A Timeless Guide to the Trek Universe,* 2002.

the most turbulent years of the civil rights movement. Apollo spacecraft were headed for the Moon, and the show went off the air the same year we first stepped foot there. By the mid-1970s, after the final Apollo mission, America was no longer heading back to the Moon, and the public needed to keep the dream, any dream, alive. With a rapidly growing baseline of support, *Star Trek* became more successful as reruns during the 1970s than it had been as a first-run show during the 1960s.

No doubt other reasons also contributed to its success. Perhaps it was the social chemistry of the international, racially integrated crew, which supplied television's first interracial kiss; or the crew's keen sense of interstellar morality when exploring alien cultures and civilizations; or the show's glimpse into our technologized spacefaring future; or the indelible split infinitive in "to boldly go where no man has gone before," spoken over the opening credits. Or maybe it was the portrayal of risk on alien planets, as the landing parties would persistently lose a crew member to unforeseen dangers.

I cannot speak for all Trekkers. Especially since I do not count myself among them, never having memorized the floor plans of the original starship Enterprise, nor donned a Klingon mask during Halloween. But as someone who, then and now, maintains a professional interest in cosmic discovery and the future technologies that will facilitate it, I offer a few reflections on the original show.

I am embarrassed to admit (don't tell anybody) that when I first saw the interior doors on the Enterprise slide open automatically as crew members walk up to them, I was certain that such a mechanism would not be invented during my years on Earth. *Star Trek* was taking place hundreds of years hence, and I was observing future technology. Same goes for those incredible pocket-size data disks they insert into talking computers. And those palm-size devices they use to talk to one another. And that square cavity in the wall that dispenses heated food in seconds. Not in my century, I thought. Not in my lifetime.

Today, obviously, we have all those technologies, and we didn't have to wait till the twenty-third century to get them. But I take pleasure in noting that our twenty-first-century communication and data-storage devices are smaller than those on *Star Trek.* And unlike their sliding doors, which make primitive whooshing sounds every time they move, our automatic doors are silent.

The most gripping episodes of the original series are those in which the solutions to challenges require a blend of logical and emotional behavior, mixed with a bit of wit and a dash of politics. These shows sample the entire range of human behavior. A persistent message to the viewer is that there's more to life than logical thinking. Even though we're watching the future, when there are no countries, no religions, and no shortages of resources, life remains complex: people (and aliens) still love and hate one another, and the thirst for power and dominance remains fully expressed across the galaxy.

Captain Kirk knows this sociopolitical landscape well, enabling him to consistently outthink, outwit, and outmaneuver the alien bad guys. Kirk's interstellar savvy also enables his legendary promiscuity with extraterrestrial women. Shapely aliens often ask Kirk, in broken English, "What is kiss?" His reply is a version of "It's an ancient human practice in which two people express how much emotion they feel for each other." And it always requires a demonstration.

Star Trek is not without its occasional gaffe. In one episode, the crew must locate a stowaway bad guy. To this end, Captain Kirk produces a clever wand that greatly enhances the sound of people's heartbeats onboard, no matter where they are hiding. While demonstrating its function to his crew, Kirk confidently declares the acoustic magnification of the device to be "one to the eleventh power." If you do the math, you get: $1 \times 1 \times 1 \times 1 \times 1 \times 1 \times 1 \times 1 \times 1 \times 1 \times 1$, which of course equals 1. I was prepared to blame William Shatner for flubbing his line, which should have been "ten to the eleventh power," except that in another episode I heard Spock make the same error, at which point I blamed the writers.

Most people, including the producers, never realized that when the starship Enterprise travels "slowly," with stars gently drifting by, its speed must still be greater than one light-year per second—or more than thirty million times the actual speed of light. If Scotty, the chief engineer, is aware of this, surely he should be declaring, "Captain, the engines can't take it."

To travel great distances quickly requires the warp drives. These are a brilliant sci-fi invention that is sufficiently based on physics to be plausible, even if technologically unforeseeable. As when you fold a sheet of paper, the warp drives bend the space between you and your destination, leaving you much closer than before. Tear a hole in the fabric of space, and you can now take a shortcut without technically exceeding the speed of light.

This trick is what allowed Captain Kirk and his Enterprise to cross the galaxy briskly—a journey that would have otherwise taken a long and boring hundred thousand years.

I have learned three of life's lessons from this series: (1) in the end, you will be judged on the integrity of your mission, whether or not your mission was successful; (2) you can always outsmart a computer; and (3) never be the first person to investigate a glowing blob of plasma on an alien planet.

Happy anniversary, *Star Trek.* Live long and prosper.

••• CHAPTER TWENTY-SIX

HOW TO PROVE YOU'VE BEEN ABDUCTED BY ALIENS*

Do I believe in UFOs or extraterrestrial visitors? Where shall I begin?

There's a fascinating frailty of the human mind that psychologists know all about, called "argument from ignorance." This is how it goes. Remember what the "U" stands for in "UFO"? You see lights flashing in the sky. You've never seen anything like this before and don't understand what it is. You say, "It's a UFO!" The "U" stands for "unidentified."

But then you say, "I don't know what it is; it must be aliens from outer space, visiting from another planet." The issue here is that if you don't know what something is, your interpretation of it should stop immediately. You don't then say it must be X or Y or Z. That's argument from ignorance. It's common. I'm not blaming anybody; it may relate to our burning need to manufacture answers because we feel uncomfortable about being steeped in ignorance.

But you can't be a scientist if you're uncomfortable with ignorance, because scientists live at the boundary between what is known and unknown in the cosmos. This is very different from the way journalists portray us. So many articles begin, "Scientists now have to go back to the drawing board." It's as though we're sitting in our offices, feet up on our desks—masters of the universe—and suddenly say, "Oops, somebody discovered something!"

---

* Adapted from Q&A segment of "Cosmic Quandaries, with Dr. Neil deGrasse Tyson," St. Petersburg College and WEDU, St. Petersburg, Florida, March 26, 2008.

No. We're always at the drawing board. If you're not at the drawing board, you're not making discoveries. You're not a scientist; you're something else. The public, on the other hand, seems to demand conclusive explanations as they leap without hesitation from statements of abject ignorance to statements of absolute certainty.

Here's something else to consider. We know—not only from research experiments in psychology but also from the history of science—that the lowest form of evidence is eyewitness testimony. Which is scary, because in a court of law it's considered one of the highest forms of evidence.

Have you all played telephone? Everybody lines up; one person starts with a story and tells it to you; you hear it and then repeat it to the next person; the next person then passes it along. What happens by the time you get to the last person, who now retells the story to everybody who's heard it already? It's completely different, right? That's because the conveyance of information has relied on eyewitness testimony—or, in this case, earwitness testimony.

So it wouldn't matter if you saw a flying saucer. In science—even with something less controversial than alien visitors, and even if you're one of my fellow scientists—when you come into my lab and say, "You've got to believe me, I saw it," I'll say, "Go home. Come back when you have some kind of evidence other than your testimony."

Human perception is rife with ways of getting things wrong. We don't like to admit it, because we have a high opinion of our biology, but it's true. Here's an example: We've all seen drawings that create optical illusions. They're lots of fun, but they should actually be called "brain failures." That's what's happening—a failure of human perception. Show us a few clever drawings, and our brains can't figure out what's going on. We're poor data-taking devices. That's why we have science; that's why we have machines. Machines don't care what side of the bed they woke up on in the morning; they don't care what they said to their spouses that day; they don't care whether they had their morning caffeine. They're emotion-free data-takers. That's what they do.

Maybe you did see visitors from another part of the galaxy. I need more than your eyewitness testimony, though. And in modern times, I need more than a photograph. Today Photoshop software probably has a UFO button. I'm not saying we haven't been visited; I'm saying the evidence brought forth thus far does not satisfy the standards of evidence that any scientist would require for any other claim.

So here's what I recommend for the next time you're abducted into a flying saucer. You're there on the slab, where of course the aliens do their sex experiments on you, and they're poking you with their instruments. Here's what you do. Yell out to the alien who's probing you, "Hey! Look over there!" And when the alien looks over there, you quickly snatch something off his shelf—an ashtray, anything—put it in your pocket, and lie back down. Then when your encounter is over and done with, you come to my lab and say, "Look what I stole from the flying saucer!" Once you bring the gizmo to the lab, the issue is no longer about eyewitness testimony, because you'll have an object of alien manufacture—and anything you pull off a flying saucer that crossed the galaxy is bound to be interesting.

Even objects produced by our own culture are interesting—like my iPhone. Not long ago, the people in power might have resurrected the witch-burning laws had I pulled this thing out. So if we could get hold of some piece of technology that had crossed the galaxy, then we could have a conversation about UFOs and extraterrestrials. Go ahead, keep trying to find them; I won't stop you. But get ready for the night you'll be abducted, because when it happens, I'll want your evidence.

Many people, including all the amateur astronomers in the world, spend a lot of time looking up. We walk out of a building, we look up. Doesn't matter what's happening, we're looking up. Yet UFO sightings are not higher among amateur astronomers than they are among the general public. In fact, they're lower. Why is that so? Because we know sky phenomena. It's what we study.

One UFO sighting in Ohio was reported by a police officer. Some people think that if you're a sheriff or a pilot or a member of the military, your testimony is somehow better than that of the average person. But everyone's testimony is bad, because we're all human. This particular police officer was tracking a light that was darting back and forth in the sky. He was chasing it in his squad car. Later it turned out that the cop was chasing the planet Venus, and that he was driving on a curved road. He was so distracted by Venus that he wasn't even conscious of turning his steering wheel back and forth.

It's yet another reminder of how feeble our sensory organs are—especially when we're confronted with unfamiliar phenomena, let alone when we're trying to describe them.

• • • CHAPTER TWENTY-SEVEN

THE FUTURE OF US SPACE TRAVEL*

Interview with Stephen Colbert, *The Colbert Report*

Stephen Colbert: This turns out to be the seventh time my next guest has been on the show. One more, and he gets a free foot-long sandwich. Please welcome Neil deGrasse Tyson. First of all, do you have your frequent guest card?

Neil deGrasse Tyson: Yes, I do.

SC: Let's talk turkey here, Neil. Barack Obama intends to cancel the Constellation program that would get us to the Moon by 2020. In his inaugural speech, he said he was going to return science to its rightful place. Does that turn out to be the dustbin of history? What's going on here, my friend?

NDT: NASA is still doing good things. That is still happening.

SC: Not with men in suits, in space.

NDT: Men in space in suits is a whole other kind of enterprise.

SC: That is science. That's what I was told when I was six.

NDT: That is *also* science. But here's what will be missing without the manned program. When you're a kid in school, who are your heroes?

---

* Adapted from an interview with Stephen Colbert, *The Colbert Report*, Comedy Central, April 8, 2010, http://www.colbertnation.com/the-colbert-report-videos/270038/april-08-2010/neil-degrasse-tyson.

SC: Not Iranian space turtles, no. Neil Armstrong! Astronauts are the supermodels of science.

NDT: Yes, they are. An astronaut is the only celebrity for whom people will line up to get their autograph without necessarily even knowing their name in advance.

SC: We're going to lose that. We're going to lose that as Americans.

NDT: There's some technology development in Obama's plan, and that's all good.

SC: Technology development: you mean robots.

NDT: Yes. I don't have a problem with that.

SC: No one wants to grow up and hear a robot land and say, "This is one small step for bleep blurt."

NDT: That's right. That would be a disappointment. However, you always want to invest in robots. The problem is, you don't want to do that to the exclusion of the rest of the manned program. I'm telling you, the manned program is what excites kids to want to become scientists in the first place.

SC: OK. Now, Obama tried to put a Band-Aid on this thing, and say, Oh, we'll still have men going to space, but we're all going to hitch a ride with the Russkies or the Europeans. If we land on Mars, how are we going to know if USA is number one if an American astronaut is standing next to a French guy? Are we going to say, "Go Earth!"? No, we're going to say, "Go USA!" Right?

NDT: I don't have a problem with hitching rides into low Earth orbit, a couple hundred miles up.

SC: That's nothing. It's child's play. I do that with a kite. That guy in the lawn chair with the balloons did that.

NDT: It's like New York to Boston. If Earth were a schoolroom globe, it would be less than half an inch above the surface.

SC: If you had this view of the Earth [turns to back wall, showing a gigantic blowup of *The Blue Marble,* a photograph of Earth as seen by the Apollo

17 astronauts on their way to the Moon in December 1972], how far away are you?

NDT: Next time you might want to display Earth with the North Pole up.

SC: We're on the side of the Moon! There's no up in space.

NDT: That's true. He's right!

SC: There's no up in space. Check and mate; I accept your apology. So, how far away are we here in the photograph?

NDT: Almost thirty thousand miles from Earth.

SC: If you can get the rocket—and the right men [points to self]—where would you send the rocket next, my friend?

NDT: I view all of space as the frontier.

SC: I view all of space as *ours*. But go ahead.

NDT: I'd like to get close and comfortable with the next asteroid that might hit us. One of them buzz-cut us just a couple of hours ago.

SC: Tonight? An asteroid took a swipe at us?

NDT: An asteroid the size of a house dipped between us and the Moon's orbit. Right about there [gesturing at the photograph]. Today!

SC: So this is war. Are we in a space war, Neil?

NDT: Kind of. But I also I want to go to Mars; a whole lot of folks want to go to Mars. There's also the Moon. You get there in three days. The next time we leave low Earth orbit, I don't want it to be a three-year journey, with people not remembering how to do it. Not since 1972 have we been more than a couple hundred miles off Earth's surface, so I want to rediscover what that's like.

SC: Neil, I share your passion for America being number one.

PART III

WHY NOT

••• CHAPTER TWENTY-EIGHT

SPACE TRAVEL TROUBLES*

Listening to space enthusiasts talk about space travel or watching blockbuster science-fiction movies might make you think that sending people to the stars is inevitable and will happen soon. Reality check: it's not and it won't—the fantasy far outstrips the facts.

A line of reasoning within the ranks of the wishful might be: "We invented flight when most people thought it was impossible. A mere sixty-five years later, we went to the Moon. It's high time we journeyed among the stars. People who say it isn't possible are ignoring history."

My rebuttal is borrowed from a legal disclaimer of the investment industry: "Past performance is not an indicator of future returns." When it comes to extracting really big money from an electorate, pure science—in this case, exploration for its own sake—doesn't rate. Yet during the 1960s, a prevailing rationale for space travel was that space was the next frontier, that we were going to the Moon because humans are innate explorers. In President Kennedy's address to a joint session of Congress on May 25, 1961, he waxed eloquent on the need for Americans to reach the next frontier. The speech included these oft-quoted lines:

> I believe that this nation should commit itself to achieving the goal, before the decade is out, of landing a man on the moon and returning him safely to the earth. No single space project in this period will be more

---

* Adapted from "Space: You Can't Get There from Here," *Natural History*, September 1998.

> impressive to mankind, or more important for the long-range exploration of space; and none will be so difficult or expensive to accomplish.

These words inspired the explorer in all of us and reverberated throughout the decade. Meanwhile, nearly every astronaut was being drawn from the military—a fact that seemed hard to reconcile with the soaring rhetoric.

A mere month before Kennedy's speech, the Soviet cosmonaut Yuri Gagarin had become the first human to be launched into Earth orbit. The Cold War was under way, the space race was on, and the Soviet Union had not yet been bested. And in fact, Kennedy did adopt a military posture in his speech to Congress, just a few paragraphs before the one quoted above. But that passage hardly ever gets cited:

> If we are to win the battle that is now going on around the world between freedom and tyranny, the dramatic achievements in space which occurred in recent weeks should have made clear to us all, as did Sputnik in 1957, the impact of this adventure on the minds of men everywhere who are attempting to make a determination of which road they should take.

Had the political landscape been different, Americans—and Congress in particular—would have been loath to part with the 4 percent of the country's budget that accomplished the task.

A trip to the Moon through the vacuum of space had been in sight, even if technologically distant, ever since 1926, when Robert Goddard perfected liquid-fuel rockets. This advance in rocketry made flight possible without the lift provided by air moving over a wing. Goddard himself realized that a trip to the Moon was finally possible but that it might be prohibitively expensive. "It might cost a million dollars," he once mused.

Calculations that were possible the day after Isaac Newton wrote down his universal law of gravitation show that an efficient trip to the Moon—in a craft escaping Earth's atmosphere at a speed of seven miles per second, and coasting the rest of the way—takes about three days. Such a trip has been taken only nine times—all of them between 1968 and 1972. Other than those nine trips, when NASA sends astronauts into "space" it launches a crew into orbit a few hundred miles above our eight-thousand-mile-diameter planet. Space travel this isn't.

What if you had told John Glenn, following his historic three orbits and successful splashdown in 1962, that thirty-seven years later NASA would send him into space again? You can bet he would never have imagined that the best we could do would be to send him back into low Earth orbit.

Space Tweets #52–#55

What if we lost the Moon? Astro-folk would be thrilled. Romantic moonlit nites wreak havoc on deep sky observing

Nov 14, 2010 1:25 PM

What if we lost the Moon? We would need to find something else upon which to blame lunatic behavior

Nov 14, 2010 1:34 PM

What if we lost the Moon? No eclipses. No moon dances. No werewolves. And Pink Floyd's album: "The Dark Side"

Nov 14, 2010 1:51 PM

What if we lost the Moon? Tides would be weak – from Sun only. And NASA might have landed humans on Mars by now

Nov 14, 2010 1:42 PM

Why all the space travel troubles?

Let's start with money. If we can send somebody to Mars for less than $100 billion, then I say let's go for it. But I've made a friendly bet with Louis Friedman, former executive director of the Planetary Society (a membership-funded organization co-founded by Carl Sagan to promote the peaceful exploration of space), that we're not going to Mars anytime soon. More specifically, in 1996 I bet him that there would be no funded plan by any government to send a manned mission to Mars during the following ten years. I had hoped to lose the bet. But the only way I could have lost was if the cost of modern missions had been brought down considerably—by a factor of ten or more—compared with those of the past.

I'm reminded of the legendary portrait of NASA's spending habits that has been making its way around the Web for a decade or so. Though some details turn out to be false, the spirit is true.* The following version was forwarded to me in the late 1990s by a Russian colleague, Oleg Gnedin:

> THE ASTRONAUT PEN
> During the heat of the space race in the 1960s, the US National Aeronautics and Space Administration decided it needed a ballpoint pen to write in the zero gravity confines of its space capsules. After considerable research and development, the Astronaut Pen was developed at a cost of approximately $1 million US. The pen worked and also enjoyed some modest success as a novelty item back here on earth. The Soviet Union, faced with the same problem, used a pencil.

Unless we have a reprise of the geopolitical circumstances that dislodged $200 billion for space travel from taxpayers' wallets in the 1960s, I will remain unconvinced that we will ever send *Homo sapiens* anywhere beyond low Earth orbit. I quote a Princeton University colleague, J. Richard Gott, who spoke on a panel a few years ago at a Hayden Planetarium symposium that touched upon the health of the manned space program: "In 1969, Wernher von Braun had a plan to send astronauts to Mars by 1982. It didn't happen. In 1989, President George [H. W.] Bush promised that we would send astronauts to Mars by the year 2019. This is not a good sign. It looks like Mars is getting farther away!"

To this I add that the most prescient prediction from the 1968 sci-fi classic *2001: A Space Odyssey* is that things can go wrong.

Space is vast and empty beyond all earthly measure. When Hollywood movies show a starship cruising through the galaxy, they typically show points of light—stars—drifting past like fireflies. But the distances between stars in a galaxy are so great that for these spaceships to move as indicated

* Before 1968 both US and Soviet astronauts relied on pencils; it was the Fisher Pen Company, not NASA, that identified the need for a "space pen," in part because of the zero-G environment but also because of the flammability of the pencil's wood and lead in the pure oxygen atmosphere of the capsule. Fisher did not bill NASA for the development costs. Nevertheless, as the truth-seeking website Snopes.com opines in "The Write Stuff," the lesson of this tale is valid, even though the example is fabricated.

would require that they travel at speeds half a billion times faster than the speed of light.

The Moon is far away compared with where you might go in a jet airplane, but it sits at the tip of our noses compared with anything else in the universe. If Earth were the size of a basketball, the Moon would be the size of a softball some ten paces away—the farthest we have ever sent people into space. On this scale, Mars at its closest is a mile away. Pluto orbits a hundred miles away. And Proxima Centauri, the star nearest to the Sun, is half a million miles away.

Let's assume money is no object. In this pretend future, our noble quest to discover new places and uncover scientific truths has become as effective as war at drumming up funds. Traveling at sufficient speed to escape not only Earth but the entire solar system—twenty-five miles per second will do—a trip to the nearest star would last a long and boring thirty thousand years. A tad too long, you say? Energy increases as the square of your speed, so if you want to double your speed you must invest four times as much energy. A tripling of your speed would require nine times as much energy. No problem. Let's just assemble some clever engineers who will build us a spaceship that can summon as much energy as we want.

How about a spaceship that travels as fast as Helios-B, the US–German solar probe that was the fastest-ever unmanned space probe? Launched in 1976, it was clocked at forty-two miles per second (more than 150,000 miles per hour) as it accelerated toward the Sun. (Note that this is only one-fiftieth of one percent of the speed of light.) Such a craft would cut the travel time to the nearest star down to a mere nineteen thousand years—nearly four times the length of recorded human history.

What we really want is a spaceship that can travel near the speed of light. How about 99 percent of light speed? All you would need is 700 million times the energy that thrust the Apollo astronauts on their way to the Moon. Actually, that's what you would need if the universe were not described by Einstein's special theory of relativity. But as Einstein correctly predicted, while your speed increases, so too does your mass, forcing you to spend even more energy to accelerate your spaceship to nearly the speed of light. A back-of-the-envelope calculation shows that you would need at least ten billion times the energy used for our Moon voyages.

No problem. Our engineers are the best. But now we learn that the closest star known to have planets is not Proxima Centauri but one that is about ten light-years away. Einstein's theory of special relativity shows that while traveling at 99 percent of the speed of light, you will age at only 14 percent the pace of everybody back on Earth, and so the round trip for you will last not twenty years but about three. On Earth, however, twenty years actually do pass by, and when you return, everyone has forgotten about you.

The Moon's distance from Earth is ten million times greater than the distance flown by the original Wright Flyer at Kitty Hawk, North Carolina. That aeroplane was designed and built by two brothers who ran a bicycle repair shop. Sixty-six years later, two Apollo 11 astronauts became the first moonwalkers. In their shop, unlike the Wright brothers', you'd find thousands of scientists and engineers building a several-hundred-million-dollar spacecraft. These are not comparable achievements. The cost and effort of space travel derive not only from the vast distances to be traveled, but also from space's supreme hostility to life.

Many will declare that early terrestrial explorers also had it bad. Consider Gonzalo Pizarro's 1540 expedition from Quito across Peru in search of the fabled land of oriental spices. Oppressive terrain and hostile natives ultimately led to the death of half of Pizarro's expedition party of more than four thousand. In his mid-nineteenth-century account of this ill-fated adventure, *History of the Conquest of Peru,* William H. Prescott describes the state of the expedition party a year into the journey:

> At every step of their way, they were obliged to hew open a passage with their axes, while their garments, rotting from the effects of the drenching rains to which they had been exposed, caught in every bush and bramble, and hung about them in shreds. Their provisions spoiled by the weather, had long since failed, and the live stock which they had taken with them had either been consumed or made their escape in the woods and mountain passes. They had set out with nearly a thousand dogs, many of them of the ferocious breed used in hunting down the unfortunate natives. These they now gladly killed, but their miserable carcasses furnished a lean banquet for the famished travelers.

On the brink of abandoning all hope, Pizarro and his men built from scratch a boat large enough to take half the remaining men along the Napo River in search of food and supplies:

> The forests furnished him with timber; the shoes of the horses which had died on the road or had been slaughtered for food, were converted into nails; gum distilled from the trees took the place of pitch; and the tattered garments of the soldiers supplied a substitute for oakum. . . . At the end of two months, a brigantine was completed, rudely put together, but strong and of sufficient burden to carry half the company.

Pizarro transferred command of the makeshift boat to Francisco de Orellana, a cavalier from Trujillo, and stayed behind to wait. After many weeks, Pizarro gave up on Orellana and returned to the town of Quito, taking yet another year to get there. Later Pizarro learned that Orellana had successfully navigated his boat down the Napo River to the Amazon and, with no intention of returning, had continued along the Amazon until he emerged in the Atlantic. Orellana and his men then sailed to Cuba, where they subsequently found safe transport back to Spain.

Does this story have any lessons for would-be star travelers? Suppose one of our spacecraft with a shipload of astronauts crash-lands on a distant, hostile planet. The astronauts survive, but the spacecraft is either totaled or broken. Problem is, hostile planets tend to be considerably more dangerous than hostile natives. The planet might not have air. And what air it does have may be toxic. If the air is not toxic, the atmospheric pressure may be a hundred times higher than on Earth. If the atmospheric pressure is tolerable, the air temperature may be 200° below zero—or 200° above zero. None of these possibilities bodes well for our astronaut explorers.

But perhaps they could survive for a while on their reserve life-support system. Meanwhile, all they would need to do is mine the planet for raw materials; build another spacecraft from scratch or repair the existing damage, which might mean having to rewire the controlling computers (using whatever spare parts can be mustered from the crash site); build a rocket-fuel factory; launch themselves into space; and then fly back home.

Delightfully delusional.

• • •

Perhaps what we should do is genetically engineer new forms of intelligent life that can survive the stress of space yet still conduct scientific experiments. Actually, such creatures have already been made in the lab. They're called robots. You don't have to feed them, they don't need life support, and they won't get upset if you don't bring them back to Earth. People, on the other hand, generally want to breathe, eat, and eventually come home.

It's probably true that no city has ever held a parade for a robot. But it's probably also true that no city has ever held a parade for an astronaut who wasn't the first (or last) to do something or go somewhere. Can you name the two Apollo 12 or Apollo 16 astronauts who walked on the Moon? Probably not. Apollo 12 was the second lunar mission. Apollo 16 was the second-to-last. But I'll bet you have a favorite picture of the cosmos taken by the orbiting robot known as the Hubble Space Telescope. I'll bet you can recall images from the rovers that have six-wheeled their way across the rocky Martian landscape. I'll further bet that you've seen some jaw-dropping images of the Jovian planets—the gas giants of the outer solar system—and their zoo of moons, images taken over the decades by the Voyager, Galileo, and Cassini space probes.

In the absence of a few hundred billion dollars in travel money, and in the presence of hostile cosmic conditions, what we need is not wishful thinking and sci-fi rhetoric inspired by a cursory reading of the history of exploration. What we need—but must wait for, and indeed may never have—is a breakthrough in our scientific understanding of the structure of the universe, so that we might exploit shortcuts through the space-time continuum, perhaps through wormholes that connect one part of the cosmos to another. Then, once again, reality will become stranger than fiction.

••• CHAPTER TWENTY-NINE

REACHING FOR THE STARS*

In the months that followed space shuttle Columbia's fatal reentry through Earth's atmosphere in February 2003, everybody became a NASA critic. After the initial shock and mourning, no end of journalists, politicians, scientists, engineers, policy analysts, and ordinary taxpayers began to debate the past, present, and future of America's presence in space.

Although I have always been interested in this subject, my tour of duty with a presidential commission on the US aerospace industry has further sharpened my senses and sensitivities. Amid the occasional new arguments on the op-ed pages and TV talk shows, the same questions roll out with every new woe in the space program: Why send people instead of robots into space? Why spend money in space when we need it here on Earth? How can we get people excited about the space program again?

Yes, excitement levels are low. But lack of enthusiasm is not apathy. In this case, the business-as-usual attitude shows that space exploration has passed seamlessly into everyday culture, so most Americans no longer even notice it. We pay attention only when something goes wrong.

In the 1960s, by contrast, space was an exotic frontier—traversed by the few, the brave, and the lucky. Every gesture NASA made toward the heavens caused a splash in the media—the surest evidence that space was still unfamiliar territory.

---

* Adapted from "Reaching for the Stars," *Natural History*, April 2003.

For many, particularly for NASA aficionados and everybody employed by the aerospace industry, the 1960s were the golden era of American space exploration. A series of space missions, each more ambitious than the one before, led to six lunar landings. We walked on the Moon, just as we said we would. Surely Mars was next. Those adventures sparked an unprecedented level of public interest in science and engineering, and inspired students at every level. What followed was a domestic boom in technology that would shape our lives for the rest of the century.

A beautiful story. But let's not fool ourselves into thinking we went to the Moon because we're pioneers or explorers or selfless discovers. We went to the Moon because Cold War politics made it the militarily expedient thing to do.

What about discovery for its own sake? Are the scientific returns on a manned mission to Mars inherently important enough to justify its costs? After all, any foreseeable mission to Mars will be long and immensely expensive. But the United States is a wealthy nation. It has the money. And the technology is imaginable. Those aren't the issues.

Expensive projects are vulnerable because they take a long time and must be sustained across changeovers in political leadership as well as through downturns in the economy. Photographs of homeless children and unemployed factory workers juxtaposed with images of astronauts frolicking on Mars make a powerful case against the continued funding of space missions.

A review of history's most ambitious projects demonstrates that only defense, the lure of economic return, and the praise of power can garner large fractions of a nation's gross domestic product. In colloquial terms, that might read: You don't want to die. You don't want to die poor. And if you're smart, you'll honor those who wield authority over you. For expensive projects that fulfill more than one of these functions, money flows like beer from a freshly tapped keg. The 44,000 miles of US interstate highways offer a crisp example. Inspired by Germany's autobahns, these roads were conceived in the Eisenhower era to move matériel and personnel for the defense of the nation. The network is also heavily used by commercial vehicles, which is why there's always money for roads.

During the shuttle program the empirical risk of death was high. With two lost shuttles out of 135 launches, an astronaut's chances of not coming home were 1.5 percent. If your chances of death were 1.5 percent every time you visited the Piggly Wiggly, you would never drive your car. To the Columbia crew, however, the return was worth that risk.

I'm proud to be part of a species whose members occasionally and willingly put their lives at risk to extend the boundaries of our collective existence. Such people were the first to see what was on the other side of the cliff face. They were the first to climb the mountain. They were the first to sail the ocean. They were the first to touch the sky. And they will be the first to land on Mars.

There may be a way to keep going places, but it involves a slight shift in what the government usually calls national defense. If science and technology can win wars, as the history of military conflict suggests, then instead of counting our smart bombs, perhaps we should be counting our smart scientists and engineers. And there is no shortage of seductive projects for them to work on:

- We should search Mars for fossils and find out why liquid water no longer runs on its surface.

- We should visit an asteroid or two, and learn how to deflect them. If one is discovered headed our way, how embarrassing it would be for us big-brained, opposable-thumbed humans to meet the same fate as *T. rex*.

- We should drill through the kilometers of ice on Jupiter's moon Europa and explore the liquid ocean below for living organisms.

- We should explore Pluto and its family of icy bodies in the outer solar system, because they hold clues to our planetary origins.

- We should probe Venus's thick atmosphere to understand why its greenhouse effect has gone awry, giving rise to a surface temperature of 900° Fahrenheit.

No part of the solar system should be beyond our reach. We should deploy both robots *and* people to get there, because, among other reasons, robots make poor field geologists. And no part of the universe should hide from our telescopes. We should launch them into orbit and give them the grandest vistas for looking back at Earth and at the rest of the solar system.

With missions and projects such as those, the United States can guarantee itself an academic pipeline bursting with the best and the brightest astrophysicists, biologists, chemists, engineers, geologists, and physicists. These people will collectively form a new kind of missile silo, filled with intellectual capital. They will be ready to come forward whenever they are called, just as the nation's best and brightest have always come forward in times of need.

For the US space program to die along with the crew of the space shuttle Columbia—because nobody is willing to write the check to keep it going—would be to move backward just by standing still.

CHAPTER THIRTY

AMERICA AND THE EMERGENT SPACE POWERS*

I was born the same week NASA was founded. A few other people were born that same year: Madonna (the second one, not the first), Michael Jackson, the artist formerly known as Prince, Michelle Pfeiffer, Sharon Stone. That was the year the Barbie doll was patented and the movie *The Blob* appeared. And it was the first year the Goddard Memorial Dinner was held: 1958.

I study the universe. It's the *second* oldest profession. People have been looking up for a long time. But as an academic, it puts me a little bit outside the "club." Yes, I've spent quality time in the aerospace community, with my service on two presidential commissions, but at heart I'm an academic. Being an academic means I don't wield power over person, place, or thing. I don't command armies; I don't lead labor unions. All I have is the power of thought.

As I look around at our troubled world, I worry. Not enough people are putting thought into what they do. Allow me to provide a few examples.

One day I was reading the newspaper—a dangerous thing to do, always—and I saw a headline complaining, "HALF OF SCHOOLS IN DISTRICT SCORED BELOW AVERAGE." Well, that's kind of what an average is! You get about half below and half above.

Here's another one. "EIGHTY PERCENT OF AIRPLANE CRASH SURVIVORS

---

* Adapted from the keynote speech for the 48th Annual Dr. Robert H. Goddard Memorial Dinner, National Space Club, Washington, DC, April 1, 2005.

LOCATED EXIT DOORS BEFORE TAKE-OFF." You might be thinking, Okay, that's a good piece of information; from now on, I'm going to notice where the exit doors are. But here's the problem with that datum: suppose 100 percent of the dead people noticed where the exit doors were. You would never know, because they're dead. This is the kind of fuzzy thinking that goes on in the world today.

I've got another example. It's often said that the state lottery is a tax on the poor, because people with low incomes spend a disproportionate amount of their money on lottery tickets. It is not a tax on the poor. It's a tax on the people who never studied mathematics.

In 2002, having spent more than three years in one residence for the first time in my life, I got called for jury duty. I show up on time, ready to serve. When we get to the voir dire, the lawyer says to me, "I see you're an astrophysicist. What's that?" I answer, "Astrophysics is the laws of physics, applied to the universe—the Big Bang, black holes, that sort of thing." Then he asks, "What do you teach at Princeton?" and I say, "I teach a class on the evaluation of evidence and the relative unreliability of eyewitness testimony." Five minutes later, I'm on the street.

A few years later, jury duty again. The judge states that the defendant is charged with possession of 1,700 milligrams of cocaine. It was found on his body, he was arrested, and he is now on trial. This time, after the Q&A is over, the judge asks us whether there are any questions we'd like to ask the court, and I say, "Yes, Your Honor. Why did you say he was in possession of 1,700 milligrams of cocaine? That equals 1.7 grams. The 'thousand' cancels with the 'milli-' and you get 1.7 grams, which is less than the weight of a dime." Again I'm out on the street.

Do we say, "I'll see you in a billion nanoseconds"? Do we say, "I live just 63,360 inches up the road"? No, we don't talk that way. That's mathematically fuzzy thinking. In this case, it might even have been intentionally fuzzied.

Another area of fuzzy thinking out there is the movement called Intelligent Design. It asserts that some things are too marvelous or too intricate to explain. The contention is that these things defy common scientific accounts for cause and effect, and so they're ascribed to an intelligent, purposeful designer. It's a slippery slope.

So let's start a movement called Stupid Design, and we'll see where that takes us. For example, what's going on with your appendix? It's much better at killing you than it is at anything else. That's definitely a stupid design. What about your pinky toenail? You can barely put nail polish on it; there's no real estate there. How about bad breath, or the fact that you breathe and drink through the same hole in your body, causing some fraction of us to choke to death every year? And here's my last one. Ready? Down there between our legs, it's like an entertainment complex in the middle of a sewage system. Who designed that?

Some people want to put warning stickers on biology textbooks, saying that the theory of evolution is just one of many theories, take it or leave it. Now, religion long predates science; it'll be here forever. That's not the issue. The problem comes when religion enters the science classroom. There's no tradition of scientists knocking down the Sunday school door, telling preachers what to teach. Scientists don't picket churches. By and large—though it may not look this way today—science and religion have achieved peaceful coexistence for quite some time. In fact, the greatest conflicts in the world are not between religion and science; they're between religion and religion.

This is not simply an academic point. Let's go back a millennium. Between A.D. 800 and A.D. 1200 the intellectual center of the Western world was Baghdad. Why? Its leaders were open to whoever wanted to think stuff up: Jews, Christians, Muslims, doubters. Everybody was granted a seat at the debating table, maximizing the exchange of ideas. Meanwhile, the written wisdom of the world was being acquired by the libraries of Baghdad and translated into Arabic. As a result, the Arabs made advances in farming, commerce, engineering, medicine, mathematics, astronomy, navigation. Do you realize that two-thirds of all the named stars in the night sky have Arabic names? If you do something first and best, you get naming rights. The Arabs got naming rights to the stars twelve hundred years ago because they charted them better than anybody had done before. They pioneered the fledgling system of Hindu numerals in the new field of algebra, itself an Arabic word—which is how the numerals came to be called "Arabic numerals." "Algorithm," another familiar word, derives from the name of the Baghdad-based mathematician who also gave us the basics of algebra.

So what happened? Historians will say that with the sack of Baghdad by Mongols in the thirteenth century, the entire nonsectarian intellectual foundation of that enterprise collapsed, along with the libraries that supported it. But if you also track the cultural and religious forces at play, you find that the influential writings of the eleventh-century Muslim scholar and theologian Al-Ghazali shaped how Islam viewed the natural world. By declaring the manipulation of numbers to be the work of the devil, and by promoting the concept of Allah's will as the cause of all natural phenomena, Ghazali unwittingly quenched scientific endeavor in the Muslim world. And it has never recovered, even to this day. From 1901 to 2010, of the 543 Nobel Prize winners in the sciences, two were Muslims. Yet Muslims comprise nearly one-fourth of the world's population.

Today among fundamentalist Christians as well as Hassidic Jews, there is a comparable absence. When societies and cultures are permeated by nonsecular philosophies, science and technology and medicine stagnate. Putting warning stickers on biology books is bad practice. But if that's how the game is to be played, why not demand warning stickers on the Bible: "Some of These Stories May Not Be True."

Spring 2001, there I was, minding my own business amid the manicured lawns of the Princeton University campus—and the phone rang. It was the White House, telling me they wanted me to join a commission to study the health of the aerospace industry. Me? I don't know how to fly an airplane. At first I was indifferent. Then I read up on the aerospace industry and realized that it had lost half a million jobs in the previous fourteen years. Something bad was going on there.

The commission's first meeting was to be at the end of September. And then came 9/11.

I live—then and now—four blocks from Ground Zero. My front windows are right there. I was supposed to go to Princeton that morning, but I had some overdue writing to finish, so I stayed home. One plane goes in; another plane goes in. At that point, how indifferent could I be? I had just lost my backyard to two airplanes. Duty called. I was a changed person: not only had the nation been attacked, so had my backyard.

I distinctly remember walking into the first meeting. There were eleven other commissioners, in a room filled with testosterone. Everybody occupied space. There was General this, and Secretary of the Navy that, and Member of Congress this. It's not as though I have no testosterone, but it's Bronx testosterone. It's the kind where, if you get into a fight on the street, you kick the guy's butt. This I-build-missile-systems testosterone is a whole other kind. Even the women on the commission had it. One had a Southern accent perfectly tuned to say, "Kiss my ass." Another one was chief aerospace analyst for Morgan Stanley; having spent her life as a Navy brat, she had the industry by the gonads.

On that commission, we went around the world to see what was influencing the situation here in America. We visited China *before* they put a man in space. I had in my head the stereotype of everybody riding bicycles, but everybody was driving Audis and Mercedes Benzes and Volkswagens. Then I went home and looked at the labels on all my stuff; half of it was already being made in China. Lots of our money is going there.

On our tour we visited the Great Wall, a military project. I looked far and wide but saw no evidence of technology, just the bricks that made the wall. But I pulled out my cell phone anyway and called my mother in New York. "Oh, Neil, you're home so soon!" It was the best connection I've ever had calling her from my cell phone. Nobody in China is going, "Can you hear me now? Can you hear me now?" But it's happening throughout the Northeast Corridor. Every time you get on Amtrak, the signal goes in and out every time you pass a tree.

So when China announced, "We're going to put somebody in orbit," sure enough, I knew it was going to happen. We all knew. China says, "We want to put somebody on the Moon," I've got no doubts. When they say they want to put somebody on Mars, I'm certain of it. The thing about Mars is, it's already red, so that could work well for Chinese marketing and public relations.

After China we visited Star City in Russia, outside Moscow. Star City is the center of the Russian space program. We all crammed into the office of the head of the center, and halfway through the morning he said, "Time for vodka." The glass was so tiny that not all of my fingers fit on it, and so my pinky stuck out. I don't think you drink vodka in Russia with your pinky sticking out. Another faux pas: I was just tasting it, not swilling it, because I'm accustomed to sipping wine. So once again, I was in the vicinity of a higher stratum of testosterone.

But the visit that really made the hair rise on the back of my neck was to Brussels, where we met with European aerospace planners and executives. They had just put out their twenty-year aeronautics vision document, plus they were working on Galileo, a satellite navigation system that competes directly with our GPS. So we were kind of worried: what happens if they finish Galileo, equip European planes with it, and announce that we have to have it to fly into European airspace? We already had an ailing industry here, and retrofitting all our airplanes just to fly there would be an unwelcome financial burden. As things stood, the Europeans could use our system for free.

So, while we were trying to understand the situation, the Europeans were sitting there looking fairly smug, especially one particular guy. I'm pretty sure our chairs were a little lower than theirs, because I remember looking up at them. Considering my torso length, I should not have been looking up. And something gelled in my head. As I said, all I have is the power of thought. And I got livid.

Why was I livid? Because we were sitting around a table talking about aerospace product as though it were soybeans—what are the trade regulations, the tariffs, the restrictions; if you do this, then we'll do that. And I'm thinking, There's something wrong here. Aerospace is a frontier of our technological prowess. If you're truly on the frontier, you don't sit at a table negotiating usage rights. You're so far ahead of everybody, you're not even worried about what they want. You just give it to them. That's the posture Americans had for most of the twentieth century. In the fifties, sixties, seventies, part of the eighties, every plane that landed in your city was made in America. From Aerolineas Argentinas to Zambian Airways, everybody flew Boeings. So I got angry—not at the guy sitting across from me, but at us. I got angry with America, because advancing is not just something you do incrementally. You need innovation as well, so that you can achieve revolutionary, not merely evolutionary, advances.

One day I want to take a day trip to Tokyo. That would be a forty-five-minute ride if we go suborbital. How come we're not doing that now? If we were, I wouldn't have been at that table with the smug guy talking about the Galileo positioning system. We would already have had a pulsar navigation system, and we just wouldn't have cared about theirs. We would have been too far ahead.

• • •

So, I'm angry that aerospace has become a bargaining commodity. Also, because I'm partly an educator, when I stand in front of eighth-graders I don't want to have to say to them, "Become an aerospace engineer so that you can build an airplane that's 20 percent more fuel efficient than the ones your parents flew on." That won't get them excited. What I need to say is, "Become an aerospace engineer so that you can design the airfoil that will be the first piloted craft in the rarefied atmosphere of Mars." "Become a biologist because we need people to look for life, not only on Mars but on Europa and elsewhere in the galaxy." "Become a chemist because we want to understand more about the elements on the Moon and the molecules in space." You put that vision out there, and my job becomes easy, because I just have to point them to it and the ambition rises up within them. The flame gets lit, and they're guided on the path.

The Bush administration's vision statement has been laid down: the Moon, Mars, and beyond. There's been some controversy at the edges, but it's fundamentally a sound vision. Not enough of the public knows or understands that. But if I were the pope of Congress, I would deliver an edict to double NASA's budget. That would take it to around $40 billion. Well, somebody else in town has a $30 billion budget: the National Institutes of Health. That's fine. They ought to have a big budget, because health matters. But most high-tech medical equipment and procedures—MRIs, PET scans, ultrasound, X-rays—work on principles discovered by physicists and are based on designs developed by engineers. So you can't just fund medicine; you have to fund the rest of what's going on. Cross-pollination is fundamental to the enterprise.

Space Tweet #56
The entire half-century budget of NASA equals the current two year budget of the US military
Jul 8, 2011 11:16 AM

What happens when you double NASA's budget? The vision becomes big; it becomes real. You attract an entire generation, and generations to follow, into science and engineering. You know and I know that all emergent markets in the twenty-first century are going to be driven by science and technology. The foundations of every future economy will require it. And what happens when you stop innovating? Everyone else catches up, your jobs go overseas, and then you cry foul: Ooohh, they're paying them less over there, and the playing field is not level. Well, stop whining and start innovating.

Let's talk about true innovation. People often ask, If you like spin-off products, why not just invest in those technologies straightaway, instead of waiting for them to happen as spin-offs? The answer: it just doesn't work that way. Let's say you're a thermodynamicist, the world's expert on heat, and I ask you to build me a better oven. You might invent a convection oven, or an oven that's more insulated or that permits easier access to its contents. But no matter how much money I give you, you will not invent a microwave oven. Because that came from another place. It came from investments in communications, in radar. The microwave oven is traceable to the war effort, not to a thermodynamicist.

That's the kind of cross-pollination that goes on all the time. And that's why futurists always get it wrong—because they take the current situation and just extrapolate. They don't see surprises. So they get the picture right for about five years into the future, and they're hopeless after ten.

I claim that space is part of our culture. You've heard complaints that nobody knows the names of the astronauts, that nobody gets excited about launches, that nobody cares anymore except people in the industry. I don't believe that for a minute. When fixing the Hubble telescope was in doubt, the loudest protests came from the public. When the space shuttle Columbia broke up on reentry, the nation stopped and mourned. We may not notice something is there, but we sure as hell notice when it's not there. That's the definition of culture.

This goes deep. Last year on July 1, the Cassini spacecraft pulled into orbit around Saturn. There was nothing scientific about it, just pulling into orbit. Yet the *Today Show* figured that was news enough to put the story in their first hour—not in the second hour, along with the recipes, but in the first twenty minutes. So they called me in. When I get there, everybody says, "Congratulations! What does this mean?" I tell them it's great, that we're going to study Saturn and its moons. Matt Lauer wants to be hard-hitting, though, so he says, "But Dr. Tyson, this is a $3.3 billion mission. Given all the problems we have in the world today, how can you justify that expenditure?" So I say, "First of all, it's $3.3 billion divided by twelve. It's a twelve-year mission. Now we have the real number: less than $300 million per year. Hmmm. $300 million. Americans spend more than that per year on lip balm."

At that moment, the camera shook. You could hear the stage and lighting people giggle. Matt had no rebuttal; he just stuttered and said, "Over to you, Katie." When I exited the building, up came a round of applause from a group of bystanders who'd been watching the show. And they all held up their ChapSticks, saying, "We want to go to Saturn!"

The penetration is deep, and it's not just among engineers. When you take a taxi ride in New York, you're in the back seat, and there's a barrier there between you and the front seat, so any conversation between you and the driver has to pass through the glass. On one of my recent rides the driver, a talkative guy who couldn't have been more than twenty-three, said to me, "Wait a minute, I think I recognize your voice. Are you an expert on the galaxy?" So I said, "Yeah, I suppose." And he said, "Wow, I saw you on a program. It was the best."

He wasn't interested in me because of celebrity. That's a different kind of encounter; that's people asking you where you live and what's your favorite color. But no. He starts asking questions: Tell me more about black holes. Tell me more about the galaxy. Tell me more about the search for life. We get to the destination, I'm ready to hand him the money, and he says, "No, keep it." This guy's twenty-three years old, with a wife and a kid at home, and he's driving a taxi. I'm trying to pay him for the ride, and he declines it. That's how excited he is that he could learn about the universe.

Here's another one. I'm walking my daughter to school, and I'm ready to cross the street with her. A garbage truck stops right in the crosswalk. Garbage trucks don't stop in crosswalks. This one stops. And I'm thinking, There was a movie where a garbage truck drove past a guy, and he wasn't there after it passed. So this worries me a little. Then the driver opens the door—never seen this man in my life—and he calls out, "Dr. Tyson, how are the planets today?" I wanted to go and kiss him.

Here's my best story of all. It happened at the Rose Center for Earth and Space, where I work. There's a janitor there who I've never seen having a conversation with anyone for the three years he's been working there. You never know who's who at these entry-level positions: maybe he's mute, maybe he's a little slow. I just don't know. And then one day, out of the blue, he stops sweeping when he catches sight of me; he stands there holding onto his broom proudly, with posture; and he says, "Dr. Tyson, I have a question. Do you have a minute?" I assume he's going to ask about the employment situation, and I say, "Yeah, sure, go ahead." Then he says, "I've been thinking. I see all these pictures from the Hubble telescope, and I see all of these gas clouds. And I learned that stars are made of gas. So could it be true that the stars were made inside those gas clouds?" This is the janitor who didn't say a word for three years, and his first sentence to me is about the astrophysics of the interstellar medium. I ran up to my office, grabbed all seven of my books, handed them to him, and said, "Here, commune with the cosmos. You need more of this."

My final quote of the day says it all: "There are lots of things I have to do to become an astronaut. But first I have to go to kindergarten."—Cyrus Corey, age four.

If you double NASA's budget, whole legions of students will fill the pipeline. Even if they don't become aerospace engineers, we will have scientifically literate people coming up through the ranks—people who might invent stuff and create the foundations of tomorrow's economy. But that's not all. Suppose the next terrorist attack is biological warfare? Who are we going to call? We want the best biologists in the world. If there's chemical warfare, we want the best chemists. And we would have them, because they'd be working on problems relating to Mars, problems relating to Europa. We would have attracted those people because the vision was in place. We wouldn't have lost them to other professions. They wouldn't have become lawyers or investment bankers, which is what happened in the 1980s and 1990s.

So this $40 billion starts looking pretty cheap. It becomes not only an investment in tomorrow's economy but an investment in our security. Our most precious asset is our enthusiasm for what we do as a nation. Marshal it. Cherish it.

CHAPTER THIRTY-ONE

DELUSIONS OF SPACE ENTHUSIASTS*

Human ingenuity seldom fails to improve on the fruits of human invention. Whatever may have dazzled everyone on its debut is almost guaranteed to be superseded and, someday, to look quaint.

In 2000 B.C. a pair of ice skates made of polished animal bone and leather thongs was a transportation breakthrough. In 1610 Galileo's eight-power telescope was an astonishing tool of detection, capable of giving the senators of Venice the power to identify hostile ships before they could enter the lagoon. In 1887 the one-horsepower Benz Patent Motorwagen was the first commercially produced car powered by an internal combustion engine. In 1946 the thirty-ton, showroom-size ENIAC, with its eighteen thousand vacuum tubes and six thousand manual switches, pioneered electronic computing.

Today you can glide across roadways on in-line skates, gaze at images of faraway galaxies brought to you by spaceborne telescopes, cruise the autobahn at 170 miles an hour in a six-hundred-horsepower roadster, and carry your three-pound, wirelessly networked laptop to an outdoor café.

Of course, such advances don't just fall from the sky. Clever people think them up. Problem is, to turn a clever idea into reality, somebody has to write the check. And when market forces shift, those somebodies may lose interest and the checks may stop coming. If computer companies had stopped innovating in 1978, your desk might still sport a hundred-

---

* Adapted from "Delusions of Space Enthusiasts," *Natural History*, November 2006.

pound IBM 5110. If communications companies had stopped innovating in 1973, you might still be schlepping a two-pound, nine-inch-long cell phone. And if in 1968 the US space industry had stopped developing bigger and better rockets to launch humans beyond the Moon, we'd never have surpassed the Saturn V rocket.

Oops!

Sorry about that. We *haven't* surpassed the Saturn V, the largest, most powerful rocket flown by anybody, ever. The thirty-six-story-tall Saturn V was the first and only rocket to launch people from Earth to someplace else in the universe; it enabled every Apollo mission to the Moon from 1969 through 1972, as well as the 1973 launch of Skylab 1, the first US space station.

Inspired in part by the successes of the Saturn V and the momentum of the Apollo program, visionaries of the day foretold a future that never came to be: space habitats, Moon bases, and Mars colonies up and running by the 1990s. But funding for the Saturn V evaporated as the Moon missions wound down. Additional production runs were canceled, the manufacturers' specialized machine tools were destroyed, and skilled personnel had to find work on other projects. Today US engineers can't even build a Saturn V clone.

What cultural forces froze the Saturn V rocket in time and space? What misconceptions led to the gap between expectation and reality?

Soothsaying tends to come in two flavors: doubt and delirium. It was doubt that led skeptics to declare that the atom would never be split, the sound barrier would never be broken, and people would never want or need computers in their homes. But in the case of the Saturn V rocket, it was delirium that misled futurists into assuming the Saturn V was an auspicious beginning—never considering that it could, instead, be an end.

Space Tweets #57 & #58

Many lament the end [of] our 30-year Space Shuttle program. But is there any technology – at all – from 1981 that you still use?

Jul 21, 2011 5:43 AM

No. Unlike the Space Shuttle, the Afro pick you still use from 1976 does not count as decades-old technology

July 25, 2011 4:58 PM

• • •

On December 30, 1900, for its last Sunday paper of the year, the *Brooklyn Daily Eagle* published a sixteen-page supplement headlined "THINGS WILL BE SO DIFFERENT A HUNDRED YEARS HENCE." The contributors—business leaders, military men, pastors, politicians, and experts of every persuasion—imagined what housework, poverty, religion, sanitation, and war would be like in the year 2000. They enthused about the potential of electricity and the automobile. There was even a map of the world-to-be, showing an American Federation comprising most of the Western Hemisphere from the lands above the Arctic Circle down to the archipelago of Tierra del Fuego—plus sub-Saharan Africa, the southern half of Australia, and all of New Zealand.

Most of the writers portrayed an expansive future. George H. Daniels, however, a man of authority at the New York Central and Hudson River Railroad, peered into his crystal ball and boneheadedly predicted:

> It is scarcely possible that the twentieth century will witness improvements in transportation that will be as great as were those of the nineteenth century.

Elsewhere in his article, Daniels envisioned affordable global tourism and the diffusion of white bread to China and Japan. Yet he simply couldn't imagine what might replace steam as the power source for ground transportation, let alone a vehicle moving through the air. Even though he stood on the doorstep of the twentieth century, this manager of the world's biggest railroad system could not see beyond the automobile, the locomotive, and the steamship.

Three years later, almost to the day, Wilbur and Orville Wright made the first-ever series of powered, controlled, heavier-than-air flights. In 1957 the USSR launched the first satellite into Earth orbit. And in 1969 two Americans became the first human beings to walk on the Moon.

Daniels is hardly the only person to have misread the technological future. Even experts who aren't totally deluded can have tunnel vision. On page 13 of the *Eagle*'s Sunday supplement, the principal examiner at the US Patent Office, W. W. Townsend, wrote, "The automobile may be the vehicle of the decade, but the air ship is the conveyance of the century." Sounds visionary, until you read further. What he was talking about were blimps and zeppelins. Both Daniels and Townsend, otherwise well-informed citizens of a changing world, were clueless about what tomorrow's technology would bring.

Even the Wright brothers were guilty of doubt about the future of aviation. In 1901, discouraged by a summer's worth of unsuccessful tests with a glider, Wilbur told Orville it would take another fifty years for someone to fly. Nope: the birth of aviation was just two years away. On the windy, chilly morning of December 17, 1903, starting from a North Carolina sand dune called Kill Devil Hill, Orville was the first to fly the brothers' six-hundred-pound plane through the air. His epochal journey lasted twelve seconds and covered 120 feet—about the distance a child can throw a ball.

Judging by what the mathematician, astronomer, and Royal Society gold medalist Simon Newcomb had published just two months earlier, the flight from Kill Devil Hill should never have taken place when it did:

> Quite likely the twentieth century is destined to see the natural forces which will enable us to fly from continent to continent with a speed far exceeding that of the bird.
>
> But when we inquire whether aerial flight is possible in the present state of our knowledge; whether, with such materials as we possess, a combination of steel, cloth and wire can be made which, moved by the power of electricity or steam, shall form a successful flying machine, the outlook may be altogether different.

Some representatives of informed public opinion went even further. The *New York Times* was steeped in doubt just one week before the Wright brothers went aloft in the original Wright Flyer. Writing on December 10, 1903—not about the Wrights but about their illustrious and publicly funded competitor, Samuel P. Langley, an astronomer, physicist, and chief administrator of the Smithsonian Institution—the *Times* declared:

> We hope that Professor Langley will not put his substantial greatness as a scientist in further peril by continuing to waste his time, and the money involved, in further airship experiments. Life is short, and he is capable of services to humanity incomparably greater than can be expected to result from trying to fly.

You might think attitudes would have changed as soon as people from several countries had made their first flights. But no. Wilbur Wright wrote in 1909 that no flying machine would ever make the journey from New York to Paris. Richard Burdon Haldane, the British secretary of war, told Parliament in 1909 that even though the airplane might one day be capable of great things, "from the war point of view, it is not so at present." Ferdinand Foch, a highly regarded French military strategist and the supreme commander of the Allied forces near the end of World War I, opined in 1911 that airplanes were interesting toys but had no military value. Late that same year, near Tripoli, an Italian plane became the first to drop a bomb.

Early attitudes about flight beyond Earth's atmosphere followed a similar trajectory. True, plenty of philosophers, scientists, and sci-fi writers had thought long and hard about outer space. The sixteenth-century philosopher-friar Giordano Bruno proposed that intelligent beings inhabited an infinitude of worlds. The seventeenth-century soldier-writer Savinien de Cyrano de Bergerac portrayed the Moon as a world with forests, violets, and people.

But those writings were fantasies, not blueprints for action. By the early twentieth century, electricity, telephones, automobiles, radios, airplanes, and countless other engineering marvels were all becoming basic features of modern life. So couldn't earthlings build machines capable of space travel? Many people who should have known better said it couldn't be done, even after the successful 1942 test launch of the world's first long-range ballistic missile, the deadly V-2 rocket. Capable of punching through Earth's atmosphere, it was a crucial step toward reaching the Moon.

Richard van der Riet Woolley, the eleventh British Astronomer Royal, is the source of a particularly woolly remark. When he landed in London after a thirty-six-hour flight from Australia, some reporters asked him about space travel. "It's utter bilge," he answered. That was in early 1956. In early 1957 Lee De Forest, a prolific American inventor who helped birth the age of electronics, declared, "Man will never reach the moon, regardless of all future scientific advances." Remember what happened in late 1957? Not just one but two Soviet Sputniks entered Earth orbit. The space race had begun.

Whenever someone says an idea is "bilge" (British for "baloney"), you must first ask whether it violates any well-tested laws of physics. If so, the idea is likely to be bilge. If not, the only challenge is to find a clever engineer—and, of course, a committed source of funding.

The day the Soviet Union launched Sputnik 1, a chapter of science fiction became science fact, and the future became the present. All of a sudden, futurists went overboard with their enthusiasm. The delusion that technology would advance at lightning speed replaced the delusion that it would barely advance at all. Experts went from having much too little confidence in the pace of technological change to having much too much. And the guiltiest people of all were the space enthusiasts.

Commentators became fond of twenty-year intervals, within which some previously inconceivable goal would supposedly be accomplished. On January 6, 1967, in a front-page story, the *Wall Street Journal* announced: "The most ambitious US space endeavor in the years ahead will be the campaign to land men on neighboring Mars. Most experts estimate the task can be accomplished by 1985." The very next month, in its debut issue, *The Futurist* magazine announced that according to long-range forecasts by the RAND Corporation, a pioneer think-tank, there was a 60 percent probability that a manned lunar base would exist by 1986. In *The Book of Predictions,* published in 1980, the rocket pioneer Robert C. Truax forecast that fifty thousand people would be living and working in space by the year 2000. When that benchmark year arrived, people were indeed living and working in space. But the tally was not fifty thousand. It was three: the first crew of the International Space Station.

All those visionaries (and countless others) never really grasped the forces that drive technological progress. In Wilbur and Orville's day, you

could tinker your way into major engineering advances. Their first airplane did not require a grant from the National Science Foundation: they funded it through their bicycle business. The brothers constructed the wings and fuselage themselves, with tools they already owned, and got their resourceful bicycle mechanic, Charles E. Taylor, to design and hand-build the engine. The operation was basically two guys and a garage.

Space exploration unfolds on an entirely different scale. The first moonwalkers were two guys, too—Neil Armstrong and Buzz Aldrin—but behind them loomed the force of a mandate from President Kennedy, ten thousand engineers, $100 billion for the Apollo program, and a Saturn V rocket.

Notwithstanding the sanitized memories so many of us have of the Apollo era, Americans were not first on the Moon because we're explorers by nature or because our country is committed to the pursuit of knowledge. We got to the Moon first because the United States was out to beat the Soviet Union, to win the Cold War any way we could. Kennedy made that clear when he complained to top NASA officials in November 1962:

> I'm not that interested in space. I think it's good, I think we ought to know about it, we're ready to spend reasonable amounts of money. But we're talking about these fantastic expenditures which wreck our budget and all these other domestic programs and the only justification for it in my opinion to do it in this time or fashion is because we hope to beat [the Soviet Union] and demonstrate that starting behind, as we did by a couple of years, by God, we passed them.

Like it or not, war (cold or hot) is the most powerful funding driver in the public arsenal. Lofty goals such as curiosity, discovery, exploration, and science can get you money for modest-size projects, provided they resonate with the political and cultural views of the moment. But big, expensive activities are inherently long term, and require sustained investment that must survive economic fluctuations and changes in the political winds.

In all eras, across time and culture, only war, greed, and the celebration of royal or religious power have fulfilled that funding requirement. Today, the power of kings is supplanted by elected governments, and the power of religion is often expressed in nonarchitectural undertakings, leaving war and greed to run the show. Sometimes those two drivers work hand in hand, as in the art of profiteering from the art of war. But war itself remains the ultimate and most compelling rationale.

I was eleven years old during the voyage of Apollo 11 and had already identified the universe as my life's passion. Unlike so many other people who watched Neil Armstrong's first steps on the Moon, I wasn't jubilant. I was simply relieved that someone was finally exploring another world. To me, Apollo 11 was clearly the beginning of an era.

But I, too, was delirious. The lunar landings continued for three and a half years. Then they stopped. The Apollo program became the end of an era, not the beginning. And as the Moon voyages receded in time and memory, they seemed ever more unreal in the history of human projects.

Unlike the first ice skates or the first airplane or the first desktop computer—artifacts that make us all chuckle when we see them today—the first rocket to the Moon, the Saturn V, elicits awe, even reverence. Saturn V relics lie in state at the Johnson Space Center in Texas, the Kennedy Space Center in Florida, and the US Space and Rocket Center in Alabama. Streams of worshippers walk the rocket's length. They touch the mighty nozzles at the base and wonder how something so large could ever have bested Earth's gravity. To transform their awe into chuckles, our country will have to resume the effort to "boldly go where no man has gone before." Only then will the Saturn V look as quaint as every other invention that human ingenuity has paid the compliment of improving upon.

••• CHAPTER THIRTY-TWO

PERCHANCE TO DREAM*

When I was asked to give the keynote address at this year's Space Technology Hall of Fame dinner, I thought it was a bit odd because I serve on the board of the Space Foundation, which is the sponsor not only of this dinner but of this entire symposium, and board members are not typically asked to give keynote addresses. But this past Tuesday, when I was asked to speak, I was assured it wasn't because someone else had canceled. So I agreed, and then looked at the list of past speakers for this event: Colonel Brewster Shaw, decorated astronaut; Colonel Fred Gregory, decorated astronaut; James Albaugh, CEO, Boeing Integrated Defense Systems; Ron Sugar, CEO, Northrop Grumman; David Thompson, CEO, Spectrum Astro; Norm Augustine, CEO of Lockheed Martin, chair of the Advisory Committee on the Future of the US Space Program, chair of half a dozen other associations and academies. Having looked at the list, I realized I would be the lowest-ranking person ever to give this keynote.

True, I've never been in the military. I'm not a general or a colonel. I'm not anything. Maybe I'm a cadet. Generals have the stars and the bars. But have you noticed my vest? I've got stars—and suns and moons and planets. That would make me a space cadet.

---

* Adapted from the keynote speech for the Space Technology Hall of Fame dinner, 23rd National Space Symposium, April 12, 2007, Colorado Springs, Colorado.

As long as I've been on the Space Foundation board, I've tried to fit in. But it's hard, because my expertise is in astrophysics, and so I hang out with academic folk. We've got our own conferences. So every year that I come to the National Space Symposium and tour the exhibit hall, I feel like an anthropologist researching a tribe. I make observations that would be obvious to anthropologists but may pass unnoticed by most of you.

For example, the generals are taller than the colonels, on average. The colonels are taller than the majors, on average. If you think about it, it should be the opposite—because if you're tall, you're a bigger target on the field. Logically, then, the higher your rank, the smaller you would be. The generals would be really little people. But that's not the case.

Space Tweet #59

Just a FYI: Within two minutes of flight, the Shuttle's air-speed exceeds that of a bullet fired from an M16 assault rifle

May 16, 2011 9:25 AM

Also, the people who staff the booths are better looking than the rest of us. I don't have a problem with that; I'm just making the observation. I know there's a sales dimension to it. But then, what's with the candy bowl? "Hey, there's a missile system I might buy—sure, give me three of those—oh, you've got bite-size Snickers! Double my order." How does that work? Do the sweets bring you more sales? Somebody should check for that. Do you get more for the M&Ms than for the Snickers? Then I thought, Well, I could be influenced by candy, because three of the booths actually had Milky Way candy bars. Now you're in my territory: the galaxy.

Here's some more anthropology: men design rockets. Even stuff that isn't rockets is designed to look like rockets. Phalluses, all of them. And I'm told that when you're testing rockets and they fail on the launchpad, euphemisms like "It was an experiment high in learning opportunities" are deployed in your press conferences. But really it's just rockets suffering from projectile dysfunction. That's what it should be called: projectile dysfunction.

So I asked myself, Would rockets look this way if women designed them? It's just a question. I don't know. But I bet I know what you're think-

ing. You're thinking, They have to be designed this way because phalluses are aerodynamic. Now, rockets in the vacuum of space don't have to be aerodynamic at all, because there's no air. So for that phase of any rocket's journey, it does not need to look like a rocket. We're together on that point.

But how about when the rocket traverses the atmosphere? I wondered whether you could have a flying object that's aerodynamic yet does not derive from a phallic fixation. After exploring the problem a little further, I found a design by Philip W. Swift that he entered in a *Scientific American* paper airplane contest in the 1960s—and here it is. Nothing phallic about it. You could even say it has an opposite design. Now watch it fly!

Neil deGrasse Tyson

Well, there's ten minutes of your life you'll never get back.

So let's talk politics. I'm an academic; I lord over nothing on the landscape of people, place, or thing. But we academics, we scientists, like to argue, because that's how the fresh ideas surface. We hash things out, find a way to do the experiment better, see what works, what doesn't. So scientists are good at looking at different points of view—which, to some people, makes us look like hypocrites. We can take one point of view one day, and another point of view the next day. But what we do is, we take the Hypocritic Oath. We take our multiple points of view, but—and this is something scientists all know as we argue—in the end there's not more than one truth. So, in fact, the conversation converges. Something you don't often get in politics.

Let me give you some examples. I was born and raised in New York

City. Politically, I'm left of liberal. That makes me really rare at this moment in the state of Colorado, perhaps as rare as a conservative Republican in New York City. In a crowd this large in New York, you'd say, "See that fellow in the bow tie over in the corner? That's the Republican in the room."

Have you noticed how the talk shows invite one liberal and one conservative, and they always just fight? I don't remember ever seeing a talk show where both sides declared at the end, "Hey, we're in full agreement," and walked out hand in hand. It never happens. So it makes me wonder about the utility of those confrontations, which forces me to look in the middle. I've been looking in the middle ever since I began serving on presidential commissions. Those commissions are bipartisan. You have to solve problems, even though there's hot air over here and hot air over there. Put those together, and it's a combustible mixture. So you make them combust, let the effluent gases dissipate, and look at what remains in the middle. What remains in the middle—that's America.

Recently I visited Disney World in Florida with my family, and we went to see the full-size, animatronic presidents of the United States. My kids, then ages ten and six, went in with me and we relearned the names of every president, from George W. right on up to George W. They're all there. While I was watching the puppets move and speak onstage, I thought to myself, These aren't Republicans or Democrats; these are presidents of the United States. While every one of them was in office, something interesting happened in America. And after they were out of office, in nearly every case, something important and lasting remained.

When you look at all the accusations people make nowadays—like, "Oh, you're just a peace-loving, liberal, antiwar Democrat"—you start to wonder what it means to put all those words together in the same phrase. We fought all of World War II under a Democratic president, and a Democratic president dropped the atom bombs. Being a liberal Democrat is not synonymous with being antiwar. Circumstances change over time. Decisions have to be made independent of your political party, decisions that affect the health and wealth of the nation. The polls tell us that George W. Bush has not historically been popular with the black community. Yet who's to say that, fifty or a hundred years from now, he won't be remembered for having appointed American blacks to the highest ranks of the cabinet? No previous president placed a black person into the ascension sequence for the presidency; it was a Republican president who did it. Then there's the perennial accusation that Republicans are anti-environment. But when was the Environmental Protection Agency started? Under President Nixon, a Republican.

So I see intersections across time. I see interplay. People are quick to criticize, and there are many reasons to do so—I understand that—but in the end, there at Disney World are all the presidents standing onstage, collectively defining our country.

I've got one more intersection for you—and this one isn't about presidents. In my professional community of astrophysicists, about 90 percent of us, plus or minus, are liberal, antiwar Democrats. Yet practically all of our detection hardware flows out of historical relationships with military hardware. And that connection goes back centuries. In the early 1600s Galileo heard about the invention of the telescope in the Netherlands—which they used for looking in people's windows—and he built one himself. Almost no one had thought to look up with the telescope, but Galileo did, and there he found the rings of Saturn, the phases of Venus, sunspots. Then he realized, Hey, this would be good for our defense system. So he demonstrated his instrument to the doges of Venice, and they ordered a supply of telescopes right then and there. Of course, they probably doubled their order when Galileo brought out the Snickers.

By the way, when I talk about looking in the middle, I don't mean compromising principles. I'm talking about finding principles that are fundamental to the identity of the nation and then rallying around them. Our presence in space embodies one of those principles.

It's been said before, but I'll say it again: Regardless of what the situation occasionally looks like, space is not fundamentally partisan. It is not even bipartisan. It is nonpartisan. Kennedy said, "Let's go to the Moon," but Nixon's signature is on the plaques our astronauts left there. The urge to explore space (or not) is historically decoupled from whether you are liberal or conservative, Democrat or Republican, left-wing or right-wing. And that's a good thing. It's a sign of what's left over in the middle after all the hot air cools down.

• • •

As Americans, we've taken certain things for granted. You don't notice this until you go somewhere else. We're always dreaming. Sometimes that's bad, because we dream unrealizable things. But most of the time it's been good. It has allowed us to think about tomorrow. Entire generations of Americans have thought about living a different future—a modern future—as no culture had done before. Computers were invented in America. Skyscrapers were born in America. It was America that not only envisioned but also invented the new and modern Tomorrow, driven by designs and innovations in science and technology.

A poor nation can't be expected to dream, because it doesn't have the resources to enable the realization of dreams. For the poor, dreaming just becomes an exercise in frustration, an unaffordable luxury. But many wealthy nations don't spend enough time looking at tomorrow either—and America needs to guard against becoming one of those. Although we still want to think about the future, we are in danger of becoming ill-equipped to make it happen.

In 2007 I gave a talk at UNESCO's Paris headquarters, at the celebration of Sputnik's fiftieth anniversary. There were four keynote speakers: one from Russia, one from India, one from the European Union, and me, from America. Naturally the Russian spoke first, because Sputnik went up first. What he talked about was what Sputnik had meant to the country—the pride, the privilege, the excitement. He talked about how that achievement infused what it was to be Russian.

Then came the representatives of India and the European Union, which don't have the historical space legacy that Russia and America do. Today, however, they're getting into space big time. What did their spokespeople talk about? Earth monitoring. India wants to learn more about the monsoons, which is completely understandable. But not once did either speaker discuss anything beyond Earth, and I thought to myself, Okay, we all love Earth, we all care about Earth. But do you want to do that to the exclusion of the rest of the universe?

Space Tweet #60
If Earth were size of a schoolroom globe, our atmosphere wouldn't be much thicker than the coat of lacquer on its surface
Apr 19, 2010 6:13 AM

The problem is, here you are looking at Earth—here's a cloud, there's a storm front—and meanwhile, there's an asteroid on the way. So you think Earth is safe until somebody else, somebody who had the foresight to look up, tells you that the asteroid's ready to take out your country, at which point you'll never have to worry again about whether a storm front is coming through.

And it's not just that asteroid we should be thinking about. We are flanked by planets that are experiments gone bad. To our left is the planet Venus, named for the goddess of love and beauty because it's so beautiful in the evening sky, the brightest thing up there. (By the way, Venus is likely to appear right after sunset, before the stars. So, just between you and me, if your wishes have not been coming true, it's because you've been wishing on a planet rather than a star.) Now, Venus is certainly beautiful in the evening sky, but it's fallen victim to a runaway greenhouse effect. It is 900° Fahrenheit on the surface of Venus, which is sometimes called our sister planet because it is about the same size and mass as Earth and has about the same surface gravity. Nine hundred degrees Fahrenheit. If you took a sixteen-inch pepperoni pizza and put it on your Venusian windowsill, it would cook in nine seconds. That's how hot it is now on Venus—a greenhouse experiment gone bad.

To our right is Mars, at one time drenched with running water. We know this because it has dry riverbeds, dry river deltas, dry meandering floodplains, dry lakebeds. Today the surface water is gone. We think it may have seeped down into permafrost, but in any case it's gone. So something bad happened on Mars, too.

And so you can't only monitor Earth to understand Earth. You can't claim to understand a sample of one. That is not science. In science, you need other things to compare with your sample; otherwise, you end up paying attention to the wrong parameters because you think they're relevant when they may actually not be. I'm not saying you shouldn't study Earth.

I'm saying that if you study Earth believing it's some isolated island in the middle of the cosmos, you are wrong. Possibly dead wrong. Fact is, we already know of an asteroid headed our way.

You know all the people out there who ask why we're spending so much money on NASA? Every time I personally hear someone say that, I ask them, "How much do you think NASA's getting? What fraction of your tax dollar do you think goes to NASA?" "Oh," they say, "ten cents, twenty cents." Sometimes they even say thirty or forty cents. And when I tell them it's not even a dime, not even a nickel, not even a penny, they say, "I didn't know that. I guess that's okay." When I tell them their half penny funded the beautiful images from the Hubble Space Telescope, the space shuttles, the International Space Station, all the scientific data from the inner and outer solar system and the research on the asteroid headed our way, they change their tune. But ignorance works its way up to people who perhaps should know better.

A principal task of Congress is to levy and spend our money. Occasionally, people muse that some or all of NASA's budget should go to heal the sick, feed the homeless, train the teachers, or engage whatever social programs beckon. Of course, we already spend money on all these things, and on countless other needs. It's this entire portfolio of spending that defines a nation's identity. I, for one, want to live in a nation that values dreaming as a dimension of that spending. Most, if not all, of those dreams spring from the premise that our discoveries will transform how we live.

Recently I had a depressing revelation. It was about firsts. The first cell phone looked like a large brick. You see it and you think, Did people actually hold this up to their ear? Remember the 1987 movie *Wall Street,* with Gordon Gekko, the rich guy, at his beach house in the Hamptons, talking on one of those phones? I remember thinking, Wow, that's cool! He can walk on the beach and speak to somebody on a portable phone! But now when I look back, all I can think is, How could anybody have ever used such a thing?

This is the evidence that we've moved on: you look at the first thing—the brick-size cell phone, the car with the little crank, the airplane that looks like a cloth-wrapped insect—and you say, "Put it in a museum. Keep that first internal-combustion-engine car behind a rope, and let me drive my Maserati down the freeway." You look at what came first, you comment on how cute and quaint it is, and you move on. That's how we should be reacting to everything that happened first. That's the guarantee and the knowledge that we have moved past it.

So why is it that every time I go to the Kennedy Space Center and walk up to the Saturn V rocket, I am still impressed by it? I look at it and touch it the way the apes touched the monolith in *2001*. And I'm not alone there, looking apelike as I stand there gawking. It's as though we're all thinking, How was this possible? How did we manage to go to the Moon? Now, if you haven't been near a Saturn V rocket lately, go check it out. It is awesome. But why am I looking at something from the 1960s and saying it's awesome? I want to be able to glance at the Saturn V rocket and say, "Isn't that quaint? Look what they did back in the 1960s. But now we've got something better."

Yes, we're now working on that problem. It's a little late, though. It should have happened back in the 1970s. But we all know it stopped; I don't have to retell that story. So if you want evidence that we're not innovating, it's when you start looking at the past, at the firsts, and start wishing we could be that good again. The day you find yourself saying, "Gosh, how did they do that?" the race is over. If we don't move things forward, the rest of the world will, leaving us to run after them, playing catch-up.

By the way, who moves things forward? The engineers, the scientists, the geeks. The people who, for most of the twentieth century, all the cool people mocked. But times have changed. Now the patron saint of geeks is the richest person in the world: Bill Gates. Do you know how rich Bill Gates is? I don't think you know, so I'm going to tell you.

I happen to have enough money so that if there's a dime lying on the sidewalk and I'm in a hurry, I won't bend down to pick it up. But if I see a quarter, I stop and get it. You can do laundry with quarters, you can put them in parking meters, plus they're big. So, even given my net worth, I'm still picking up quarters—but not dimes. So let's do a ratio of my net worth and what I don't pick up to Bill Gates's net worth and what he won't pick up. How little would have to be lying in the street for Bill Gates to feel it wasn't worth bothering to pick up? Forty-five thousand dollars.

You know that passage in the Bible that says, "And the meek shall inherit the Earth"? Always wondered if that was mistranslated. Perhaps it actually says, "And the geek shall inherit the Earth."

I want to get back to what it means to dream, to have a vision. To study space, you have to ask certain questions that require new kinds of cross-pollination among multiple fields. Right now I'm looking for life on Mars. I need a biologist to help me. If there's some kind of odd life on the surface, I might step on it, so bring in the biologist. If the life exists below the soils, bring in the geologist. If there's an issue with the pH of the soil, bring in the chemist. If I want to build a structure in orbit, I need to bring in the mechanical and aerospace engineers.

Today we're all under the same tent, and we're all speaking to one another. Today we realize that space is not simply an emotional frontier; it is the frontier of all the sciences. So when I stand in front of a middle-school class, I have to be able to say, "Become an aerospace engineer because we're doing amazing science out here on the frontier."

You already know this. I'm preaching to the choir here. That's why I'm proud to be part of this Space Technology Hall of Fame family. If you're going to attract the next generation, you need and want to be working on something big, something worth dreaming about, because it's what defines who we are.

Space Tweet #61
If the surviving Chilean miners are heroes (rather than victims) then what do you call the NASA & Chilean engineers who saved them?
Oct 17, 2010 7:47 AM

Maybe you're worried about scientific literacy. China has more scientifically literate people than America has college graduates. What can be

done about that? How do you attract people? I don't know a bigger force of attraction than the universe magnet. I don't twist newscasters' arms or tell them, "Do thus-and-such story on the universe tonight." I sit in my office, minding my own business, and the phone rings—because the universe flinched the day before, and they want a sound bite on it. I'm responding to an appetite that's already there. So the issue is, do we have the drive and the will to feed that appetite?

Wherever I travel, if strangers recognize me in the street, seven in ten of them are working-class. I think of them as blue-collar intellectuals. These are the people who, owing to whatever circumstance or turn of luck, could not or did not go to college. Yet they have stayed intellectually curious their entire lives. So they watch the Discovery Channel; they watch National Geographic, they watch *NOVA*; they want to know the answers. And we need to harness their desire for answers so that it helps transform the nation.

The legacy being built by the Space Technology Hall of Fame is just the beginning. I also want us to take what I call the cosmic perspective. It's the perspective that we can dream beyond ourselves, beyond Earth, that we can imagine a tomorrow that's different from today. We may not realize how rare and how privileged it is to have thoughts of tomorrow, and so I just want to ensure—through the kinds of inventions you have created, through the proper funding of programs well into the future—that we bequeath to our next generation the right and the privilege to dream. Because without that, what are we? I look at the last several decades, at how they dreamed back then and how we surfed along afterward, and I think: No. We're too powerful; we're too smart; we have too many ambitious people to deny our next generation the privilege of inventing tomorrow. And so, may none of us ever take the power of the dream for granted.

••• CHAPTER THIRTY-THREE

BY THE NUMBERS*

We've got challenges ahead of us. They're bigger than you might think. They're more severe than you might think. Recently I was invited to serve on a committee for ABC's *Good Morning America*. Our task was to pick a new set of the seven wonders of the world. Why not? It's the twenty-first century; let's do it. The resulting program would reveal one wonder of the world per day—kind of like a striptease lasting seven days.

The original seven wonders of the world were manmade things, but for our exercise, natural objects were allowed on the list. The eight others on the selection committee had traveled the world, and out came a familiar list of nature's suspects, including the Great Barrier Reef in Australia and the Amazon River basin. My suggestion was the Saturn V rocket. Hello! The Saturn V, first rocket ever to escape Earth.

When I mentioned this, they all turned and looked at me like I had three heads. I had to be polite, because we were being filmed, and so I gave my most impassioned plea: Saturn V was the first rocket to leave low Earth orbit at escape velocity—25,000 miles an hour, seven miles a second. No other spaceship had ever taken humans to that speed. The crowning achievement of human engineering and ingenuity. And once again, they all looked at me. I was not connecting. I was not communicat-

* Adapted from the keynote speech for the Space Technology Hall of Fame dinner, 24th National Space Symposium, April 10, 2008, Colorado Springs, Colorado.

ing. But the conversation sparkled when canyons, waterfalls, and ice caps were being discussed.

Then I thought, Well, let me try another plan, and I mentioned the Three Gorges Dam in China, the largest engineering project in the world, six times larger than the Hoover Dam. That category, by the way, is no stranger to China; they've had the largest engineering project in the world before. The Great Wall of China was just such a plan. So they know about big projects. The other people on the committee again turned and looked at me like I had three heads, and said, "Don't you know the dam is devastating to the environment?" I replied, "It wasn't a prerequisite that no humans would be harmed in the making of these seven wonders. And in any case, that doesn't make the largest damn dam in the world any less of an engineering marvel."

I got outvoted on that one too.

Several months later I was invited back for another round: to help pick the seven wonders of the United States of America. If I couldn't get the Saturn V listed as one of our own seven wonders, I told myself, I would just pack up and move to another country—or another planet. And yes, after some arm twisting, aggressive posturing, and strategic horse-trading, I succeeded.

But this tells us that the population is simply not plugged into what we—the space enthusiasts, the space technologists, the space visionaries—are doing. Most of what we take for granted—what we know to be the value of this enterprise to the security, the financial health, and the dreams of the nation—goes unnoticed by the public that derives daily benefits from the enterprise.

Not only that, some of them even celebrate their science illiteracy. They're not even embarrassed by it. You've been to cocktail parties where the humanities types are standing in a corner chatting about Shakespeare or Salman Rushdie or the latest Man Booker Prize winner. But if a science geek joins them and happens to mention a quick mental calculation, the most common response is, "I was never good at math," followed by a collective chuckle. Now suppose you're one of those humanities types, and you visit the geek corner and mention some aspect of grammar. Do you think the geeks will say, "Oh, I was never good at nouns and verbs"? Of course not. Whether or not they liked their English classes, they would never chuckle about being bad at the language. So I see a profound inequality in what is and isn't accepted in our collective ignorance.

I'm concerned about this kind of illiteracy. First of all, as you know, there are two kinds of people in the world: those who divide everyone into two kinds of people and those who don't. But actually, there are three kinds of people in the world: those who are good at math and those who aren't.

Our nation is turning into an idiocracy. For example, many people don't seem to grasp what an average is: half below and half above. Not all children can be above average. And why is it that three-quarters of all high-rise buildings—I've studied this—go directly from a twelfth to a fourteenth floor? Check out their elevators. Here we are in twenty-first-century America, and people who walk among us fear the number thirteen. What kind of country are we turning into? What's next—people calculating averages for things that don't average? In a statement that's arithmetically accurate yet biologically meaningless, the Irish mathematician and satirist Des MacHale noted that the average person walks around with one breast and one testicle.

The problem isn't just math. You know there's something wrong out there when you read the label on a bottle of Formula 409 Cleaner and it says, "Do not use on contact lenses." That warning can be there only because someone tried it. As the comedian Sarge notes in his act, Formula 409 gets scuff marks off linoleum. If you use it to clean your contact lenses, you're too dumb to feel it burning.

Recently I gave a talk in Saint Petersburg, Florida. The last question of the night—I don't know if this person was particularly worried about the upcoming election—was, "What would you do if, a year from now, all the money for science and engineering research was cut to zero, yet Congress allowed you to pick one project you could do? What would that project be?" I promptly replied, "I would take that money, build a ship, and sail to some other country that values investment in science. And in my rearview mirror would be all of America moving back into the caves, because that's what happens when you don't invest in science and engineering."

There was a day when Americans would construct the tallest buildings, the longest suspension bridges, the longest tunnels, the biggest dams. You might say, "Well, those are just bragging rights." Yes, they were bragging rights. But more important, they embodied a missio… working on the frontier—the technological frontier, the en… tier, the intellectual frontier—about going places that had not … the day before. When that stops, your infrastructure crumbles.

There's a lot of talk about China these days. So let's talk more ab… it. We keep hearing about ancient Chinese remedies and ancient Chinese inventions. But when do you hear about modern Chinese inventions? Here are some of the things that the Chinese achieved between the late sixth and late fifteenth centuries A.D.: They discovered the solar wind and magnetic declination. They invented matches, chess, and playing cards. They figured out that you can diagnose diabetes by analyzing urine. They invented the first mechanical clock, movable type, paper money, and the segmented-arch bridge. They basically invented the compass and showed that magnetic north is not the same as geographic north—a good thing to know when you're trying to navigate. They invented phosphorescent paint, gunpowder, flares, and fireworks. They even invented grenades. They were hugely active in international trade over that period, discovering new lands and new peoples.

And then, in the late 1400s, China turned insular. It stopped looking beyond its shores. It stopped exploring beyond its then-current state of knowledge. And the entire enterprise of creativity stopped. That's why you don't hear people saying, "Here's a modern Chinese answer to that problem." Instead they're talking about ancient Chinese remedies. There's a cost when you stop innovating and stop investing and stop exploring. That cost is severe. And it worries me deeply, because if you don't explore, you recede into irrelevance as other nations figure out the value of exploration.

What else do we know about China? It has nearly 1.5 billion people—one-fifth of the world's population. Do you know how big a billion is? In China it means that if you're one in a million, there are 1,500 other people just like you.

Not only that, the upper quartile of China—the smartest 25 percent—outnumbers the entire population of the United States. Lose sleep over that one. You've seen the numbers: China graduates about half a million scientists and engineers a year; we graduate about seventy thousand—much less than the ratio of our populations would indicate. A talk-show host in Salt Lake City recently asked me about those numbers, and I said, "Well, we raduate half a million of something a year: lawyers." So the guy asked me what that says about America, and I said, "It tells me we are going into the future fully prepared to litigate over the crumbling of our infrastructure." That's what the future of America will be.

Am I making this up about the infrastructure crumbling? No. In July 2007 a steam pipe blew up in Manhattan; people were injured; people died. The following month an eight-lane bridge over the Mississippi River, on I-35, collapsed in Minneapolis. In 2005 levees in New Orleans broke. What is going on? This is what happens when you move from being a technological leader in the world to becoming an idiocracy. Your infrastructure begins to crumble, and you just run behind the problems, trying to fix them after the damage occurs.

I don't want to build shelters to house people when a levee breaks; let's build levees that don't break in the first place. I don't want to escape from a tornado; let's figure out a way to stop the tornado. I don't want to run away from an incoming asteroid; let's figure out how to deflect it. These are two different mentalities. One of them cowers in the presence of a problem; the other solves the problem before it wreaks havoc. And the people who solve infrastructure problems are the scientists and the engineers. I'm tired of building shelters from things we could have prevented from happening.

We're listening to each other, but is anybody else listening? I don't know.

How many space people are there anyway? How many employees does Boeing have? 150,000 worldwide. Lockheed Martin: 125,000. Northrop Grumman: 120,000. General Dynamics: 90,000. NASA: 18,000. Not all of the people at those big companies are involved in space, of course, plus there are other companies with many fewer employees. How about membership organizations? The Planetary Society, the National Space Society, and the Mars Society combined: maybe 100,000 people. If you add them all up—I did this exercise—there are no more than half a million engaged in this industry in the United States. Half a million. That's one-sixth of one percent of the nation's population.

Now, here's the problem. We get viewed as though we're some kind of special interest group, so let's compare ourselves with other special interest groups. How about the NRA? More than four million members. Who's got a million members, twice as many as all the Americans who work in the aerospace industry? The Hannah Montana Fan Club. The Benevolent & Protective Order of Elks of the USA. The Arbor Day Foundation. A million children are home-schooled in America. A million people belong to gangs in America. As far as special interests go, we're way down on the list of groups to pay attention to—unless we can get the message out that what we do is fundamental to the identity of America.

Let's talk budgets for a minute. I like talking about budgets. NASA's budget, depending on which year you're talking about, is about half a penny on the tax dollar.

Space Tweet #62
The US bank bailout exceeded the half-century lifetime budget of NASA
Jul 8, 2011 11:10 AM

Many people try to justify NASA by its spin-offs—although I think we've finally let go of the Tang reference. Of course we've got spin-offs, as every year's inductees to the Space Technology Hall of Fame testify. NASA also exerts direct and indirect economic impact in every community where it does business. Its presence has fostered educated communities. Meanwhile, salaries get paid. Goods and services get purchased. Sum up the economic impact, and NASA is net positive. Yet none of this fully captures the soul of NASA's mission.

Something else captures it, though, something that's rarely talked about: the sheer joy of exploration and discovery. Not all countries offer their citizens this possibility. People living in poor countries are reduced to the three biological imperatives: the search for food, shelter, and sex. Ignore those basic requirements, and you'll go extinct. But in wealthy nations, we can go beyond the basics. We have time to reflect on our place in the cosmos. We might think of this as a luxury, but it's not. The way I see it, exploration and discovery fully express the biological imperative of our brain. To deny these yearnings is a travesty of nature.

Space knowledge is one of the fruits of using our brain. So are numbers. I like numbers, especially big numbers. I don't think most people have a feeling for how big the big numbers are. What do we call things that are big? We call them astronomical: astronomical debt, astronomical salaries. The universe deals in big numbers, and I want to share some of them with you.

Let's start out small, just to get warmed up. How about the number "1"? We understand the number "1." Go up by a power of a thousand, and we get to 1,000. That's another number we understand. Go up by another power of a thousand, and we get to 1,000,000. A million. Now we're getting to the populations of large cities. Eight of those live in New York City. Eight million people. Go up by another power of a thousand, and you get to 1,000,000,000. A billion. You know how big a billion is? I'm going to tell you.

Space Tweet #63
What country do I live in? TimeWarner Cable. @TWCable_NYC: 750 channels. (Dozens in foreign languages.) None of them NASA-TV
Feb 24, 2011 11:01 AM

McDonald's has sold a lot of hamburgers, so many that they've lost count. Just between friends, let's call it 100,000,000,000—a hundred billion. Do you know how many hamburgers that is? If you start in Colorado Springs and lay them end to end going due west, you'll get to Los Angeles, float across the Pacific, get to Japan, go across Asia and Europe and the Atlantic Ocean, come back to Washington, DC, and keep going. You'll get right back to Colorado Springs on your 100,000,000,000 hamburgers—fifty-two times over, in fact. By the way, I did this calculation based on the bun. It's a bun calculation: fifty-two times around the planet. By itself, the patty won't stretch as far. Then if you want to stack the leftover burgers, you can make a stack high enough to reach the Moon and back. That's a hundred billion for you.

Back to a billion. Anybody out there who's thirty-one years old? In this year of your life, you'll live your billionth second. It's the second that follows 259 days, one hour, forty-six minutes, and forty seconds (minus, of course, all the leap days and leap seconds of your life.) Most people celebrate their birthday. I celebrated my birth second—my billionth second—with a bottle of champagne. I'd be happy to recommend some champagne for the occasion. But you'll have to drink it real quick, because you've got only one second to celebrate.

Let's go up by another power of a thousand, to a trillion: 1,000,000,000,000. A "1" with twelve zeroes. You cannot count to a trillion. If you counted one number per second, as I just mentioned, it would take you thirty-one years to count to a billion. How long would it take you to count to a trillion? A thousand times longer—thirty-one thousand years. So don't even try it. Thirty-one thousand years ago, cave dwellers were making rock art in Australia and carving small, thick-thighed female figurines in Central Europe.

Now go up another power of a thousand, to the "1" with fifteen zeroes. Now we're at quadrillion. The estimated number of sounds and words ever uttered by all humans who have ever lived is a hundred quadrillion. That includes Congressional filibusters. They're part of the tally.

Up another power of a thousand: "1" with eighteen zeroes. That's quintillion, the average number of grains of sand on a beach—even the sand that comes home with you in your bathing suit. I counted that too.

Up yet another factor of a thousand: "1" with twenty-one zeroes. That is the number of stars in the observable universe. Sextillion stars. If you came in here with a big ego, it won't play well with that number. Consider our neighbor, the Andromeda galaxy, which is kind of like a twin of ours; within its fuzzy cloud system is the puddled light of hundreds of billions of stars. When you look farther, courtesy of the Hubble Space Telescope, you see nothing but these systems, every single one of them appearing as a smudge. Every smudge is a full red-blooded galaxy, kin to Andromeda, containing its own hundreds of billions of stars. Getting a taste of cosmic scale makes you feel small only if your ego is unjustifiably large to begin with.

In all of these galaxies, there are stars of a particular kind that manufacture heavy elements in their core and then explode, spreading their enriched contents across the galaxy—carbon, nitrogen, oxygen, silicon, and on down the periodic table of elements. These elements enrich the gas clouds that birth the next generation of stars and their associated planets, and on those planets are the ingredients of life itself, which match, one for one, the ingredients of the universe.

The number-one element in the universe is hydrogen; so, too, it is number one in the human body. Among other places, you find it in the water molecule, H_2O. Next most common in the universe is helium: chemically inert, and thus not useful to the human body. Inhaling it makes a good party trick, but it's not chemically useful to life. Next on the cosmic list is oxygen; next in the human body and all life on Earth is oxygen. Carbon comes next in the universe; carbon comes next in life. It's a hugely fertile element. We ourselves are carbon-based life. Next in the universe? Nitrogen. Next in life on Earth? Nitrogen. It all matches one for one. If we were made of an isotope of bismuth, you'd have an argument that we're something unique in the cosmos, because that would be a really rare thing to be made of. But we're not. We're made of the commonest ingredients. And that gives me a sense of belonging to the universe, a sense of participation.

Space Tweets #64 & #65

FYI: More than 90% of atoms in the universe are Hydrogen – with a single proton in its nucleus

Jul 2, 2010 9:07 AM

I remember SciFi story: Aliens crossed Galaxy to suck H from Earth's H2O supply. Author badly needed Astro101

Jul 2, 2010 9:13 AM

You could also ask who's in charge. Lots of people think, well, we're humans; we're the most intelligent and accomplished species; we're in charge. Bacteria may have a different outlook: more bacteria live and work in one linear centimeter of your lower colon than all the humans who have ever lived. That's what's going on in your digestive tract right now. Are we in charge, or are we simply hosts for bacteria? It all depends on your outlook.

I think about human intelligence a lot, because I'm worried about this idiocracy problem. But look at our DNA. It's 98+ percent identical to that of a chimpanzee, and only slightly less similar to that of other mammals. We consider ourselves smart: we compose poetry, we write music, we solve equations, we build airplanes. That's what smart creatures do. Fine. I don have a problem with that self-serving definition. I think we can agree that no matter how hard you try, you will never teach trigonometry to a chimpanzee. The chimp probably couldn't even learn the times table. Meanwhile, humans have sent spaceships to the Moon.

In other words, what we celebrate as our intelligence derives from a less than 2 percent difference in DNA. So here's a night thought to disturb your slumber. Since a genetic difference of 2 percent is so small, maybe the actual difference in intelligence is also small, and we're just ego-servingly telling ourselves it's large. Imagine a creature—another life-form on Earth, an alien, whatever—whose DNA is 2 percent beyond ours on the intelligence scale, as ours is beyond the chimp's. In that creature's presence, we would be blithering idiots.

I worry that some problems in the universe might be just too hard for the human brain. Maybe we're simply too stupid.

Some people are upset by this. Don't be. There's another way to look at it. It's not as though we're down here on Earth and the rest of the universe is out there. To begin with, we're genetically connected to each other and to all other life-forms on Earth. We're mutual participants in the biosphere. We're also chemically connected to all the other life-forms we have yet to discover. They, too, would use the same elements we find in our periodic table. They do not and cannot have some other periodic table. So we're genetically connected to each other; we're molecularly connected to other objects in the universe; and we're atomically connected to all matter in the cosmos.

For me, that is a profound thought. It is even spiritual. Science, enabled by engineering, empowered by NASA, tells us not only that we are in the universe but that the universe is in us. And for me, that sense of belonging elevates, not denigrates, the ego.

This is an epic journey my colleagues and I have been on—in my case since I was nine years old. The rest of the world needs to understand this journey. It's fundamental to our lives, to our security, to our self-image, and to our capacity to dream.

CHAPTER THIRTY-FOUR

ODE TO CHALLENGER, 1986*

Eager and ready you stood
In stately pre-launch repose.
At "Main engine start 3–2–1,"
From a mighty cloud you rose.

Your rockets thrust you skyward
But on "Throttle up" they failed.
A fireball consumed you,
Wayward boosters left their trail.

The Atlantic was below
Where Columbus first set sail.
An enterprising journey,
Where the brave alone prevailed.

Your astronauts showed courage.
With you they fell to sea.
There was pilot Michael Smith
And commander Dick Scobee.

---

* Adapted from an unpublished ode written in 1986. Editor's Note: The ode invokes words related to the names of all five space shuttles that existed in 1986—Atlantis, Challenger, Columbia, Discovery, Enterprise.

The engineers Greg Jarvis
And Judith Resnik were there;
Ellison Onizuka
And physicist Ron McNair.

Who could forget the teacher,
Christa McAuliffe? She gives
Children dreams and parents hope.
In life she died, but now lives.

Our urge to explore remains
Deep within us, 'til last breath.
But therein lies the challenge:
To discover, we risk death.

The nation stopped; the world mourned.
To space you did not climb.
Lost to NASA forever,
Hallowed forever in time.

••• CHAPTER THIRTY-FIVE

SPACECRAFT BEHAVING BADLY*

There's no sweeping it under the rug. NASA's twin Pioneer 10 and Pioneer 11 space probes, launched in the early 1970s and headed for stars in the depths of our galaxy, are both experiencing a mysterious force that has altered their expected trajectories. They're as much as a quarter-million miles closer to the Sun than they were expected to be.

That mismatch, known as the Pioneer anomaly, first became evident in the early 1980s, by which time the spacecraft were so far from the Sun that the slight outward pressure of sunlight no longer exerted significant influence over their velocity. Scientists expected that Newtonian gravity alone—traceable to the Sun and all that orbits it—would thenceforth account for the pace of the Pioneers' journey. But things seemingly haven't turned out that way. The extra little push from solar radiation had been masking an anomaly. Once the Pioneers reached the point where the sunlight's influence was less than the anomaly's, both spacecraft began to register an unexplained, persistent change in velocity—a sunward force, a drag—operating at the rate of a couple hundred-millionths of an inch per second for every second of time the twins have been traveling. That may not sound like much, but it eventually claimed thousands of miles of lost ground for every year out on the road.

Contrary to stereotype, research scientists don't sit around their offices smugly celebrating their mastery of cosmic truths. Nor are scientific dis-

* Adapted from "Spacecraft Behaving Badly," *Natural History*, April 2008.

coveries normally heralded by people in lab coats proclaiming, "Eureka!" Instead, researchers say things like, "Hmm, that's odd." From such humble beginnings come mostly dead ends and frustration, but also an occasional new insight into the laws of the universe.

And so, once the Pioneer anomaly revealed itself, scientists (predictably) said, "Hmm, that's odd." They kept looking, and the oddness didn't go away. Serious investigation began in 1994, the first research paper about it appeared in 1998, and since then all sorts of explanations have been proffered to account for the anomaly. Contenders that have now been ruled out include software bugs, leaky valves in the midcourse-correction rockets, the solar wind interacting with the probes' radio signals, the probes' magnetic fields interacting with the Sun's magnetic field, the gravity exerted by newly discovered Kuiper Belt objects, the deformability of space and time, and the accelerating expansion of the universe. The remaining explanations range from the everyday to the exotic. Among them is the suspicion that in the outer solar system, Newtonian gravity begins to fail.

The very first spacecraft in the Pioneer program—Pioneer 0 (that's right, "zero")—was launched, unsuccessfully, in the summer of 1958. Fourteen more were launched over the next two decades. Pioneers 3 and 4 studied the Moon; 5 through 9 monitored the Sun; 10 flew by Jupiter; 11 flew by Jupiter and Saturn; 12 and 13 visited Venus.

Pioneer 10 left Cape Canaveral on the evening of March 2, 1972—nine months before the Apollo program's final Moon landing—and crossed the Moon's orbit the very next morning. In July 1972 it became the first human-made object to traverse the asteroid belt, the band of rocky rubble that separates the inner solar system from the giant outer planets. In December 1973 it became the first to get a "gravity assist" from massive Jupiter, which helped kick it out of the solar system for good. Although NASA planned for Pioneer 10 to keep signaling Earth for a mere twenty-one months, the craft's power sources kept going and going—enabling the fellow to call home for thirty years, until January 22, 2003. Its twin, Pioneer 11, had a shorter signaling life, with its final transmission arriving on September 30, 1995.

At the heart of Pioneers 10 and 11 is a toolbox-size equipment compartment, from which booms holding instruments and a miniature power plant project at various angles. More instruments and several antennas are clamped to the compartment itself. Heat-responsive louvers keep the onboard electronics at ideal operating temperatures, and there are three pairs of rocket thrusters, packed with reliable propellant, designed to provide midcourse corrections en route to Jupiter.

Power for the twins and their fifteen scientific instruments comes from radioactive chunks of plutonium-238, which drive four radioisotope thermoelectric generators. The heat from the slowly decaying plutonium, with its half-life of eighty-eight years, yielded enough electricity to run the spacecraft, photograph Jupiter and its satellites in multiple wavelengths, record sundry cosmic phenomena, and conduct experiments more or less continuously for upward of a decade. But by April 2001 the signal from Pioneer 10 had dwindled to a barely detectable billionth of a trillionth of a watt.

The probes' main agent of communication is a nine-foot-wide, dish-shaped antenna pointed toward Earth. To preserve the antenna's alignment, each spacecraft has star and Sun sensors that keep it spinning along the antenna's central axis in much the way that a quarterback spins a football around its long axis to stabilize the ball's trajectory. For the duration of the dish antenna's prolonged life, it sent and received radio signals via the Deep Space Network, an ensemble of sensitive antennas that span the globe, making it possible for engineers to monitor the spacecraft without a moment's interruption.

The famous finishing touch on Pioneers 10 and 11 is a gold-plated plaque affixed to the side of the craft. The plaque includes an engraved illustration of a naked adult male and female; a sketch of the spacecraft itself, shown in correct proportion to the humans; and a diagram of the Sun's position in the Milky Way, announcing the spacecraft's provenance to any intelligent aliens who might stumble across one of the twins. I've always had my doubts about this cosmic calling card. Most people wouldn't give their home address to a stranger in the street, even when the stranger is one of our own species. Why, then, give our home address to aliens from another planet?

• • •

Space travel involves a lot of coasting. Typically, a spacecraft relies on rockets to get itself off the ground and on its way. Other, smaller engines may fire en route to refine the craft's trajectory or pull the craft into orbit around a target object. In between, it simply coasts. For engineers to calculate a craft's Newtonian trajectory between any two points in the solar system, they must account for every single source of gravity along the way, including comets, asteroids, moons, and planets. As an added challenge, they must aim for where the target should be when the spacecraft is due to arrive, not for the target's current location.

Calculations completed, off went Pioneers 10 and 11 on their multi-billion-mile journeys through interplanetary space—boldly going where no hardware had gone before, and opening new vistas on the planets of our solar system. Little did anyone foresee that in their twilight years the twins would also become unwitting probes of the fundamental laws of gravitational physics.

Astrophysicists do not normally discover new laws of nature. We cannot manipulate the objects of our scrutiny. Our telescopes are passive probes that cannot tell the cosmos what to do. Yet they can tell us when something isn't following orders. Take the planet Uranus, whose discovery is credited to the English astronomer William Herschel and dated to 1781 (others had already noted its presence in the sky but misidentified it as a star). As observational data about its orbit accumulated over the following decades, people began to notice that Uranus deviated slightly from the dictates of Newton's laws of gravity, which by then had withstood a century's worth of testing on the other planets and their moons. Some prominent astronomers suggested that perhaps Newton's laws begin to break down at such great distances from the Sun.

Space Tweet #66

Isaac Newton: Smartest ever. Discovered laws of motion gravity & optics. Invented calculus in spare time. Then turned 26

May 14, 2010 3:18 AM

What to do? Abandon or modify Newton's laws and dream up new rules of gravity? Or postulate a yet-to-be-discovered planet in the outer solar system, whose gravity was absent from the calculations for Uranus's orbit? The answer came in 1846, when astronomers discovered the planet Neptune just where a planet had to be for its gravity to perturb Uranus in just the ways measured. Newton's laws were safe . . . for the time being.

Then there's Mercury, the planet closest to the Sun. Its orbit, too, habitually disobeyed Newton's laws of gravity. Having predicted Neptune's position on the sky within one degree, the French astronomer Urbain-Jean-Joseph Le Verrier now postulated two possible causes for Mercury's deviant behavior. Either it was another new planet (call it Vulcan) orbiting so close to the Sun that it would be well-nigh impossible to discover in the solar glare, or it was an entire, uncatalogued belt of asteroids orbiting between Mercury and the Sun.

Turns out Le Verrier was wrong on both counts. This time he really did need a new understanding of gravity. Within the limits of precision that our measuring tools impose, Newton's laws behave well in the outer solar system. However, they break down in the inner solar system, where they are superseded by Einstein's general relativity. The closer you are to the Sun, the less you can ignore the exotic effects of its powerful gravitational field.

Two planets. Two similar-looking anomalies. Two completely different explanations.

Pioneer 10 had been coasting through space for less than a decade and was around 15 AU from the Sun when John D. Anderson, a specialist in celestial mechanics and radio-wave physics at NASA's Jet Propulsion Laboratory (JPL), first noticed that the data were drifting away from the predictions made by JPL's computer model. (One AU, or astronomical unit, represents the average distance between Earth and the Sun; it's a "yardstick" for measuring distances within the solar system.) By the time Pioneer 10 reached 20 AU, a distance at which pressure from the Sun's rays no longer mattered much to the trajectory of the spacecraft, the drift was unmistakable. Initially Anderson didn't fuss over the discrepancy, thinking the problem could probably be blamed on either the software or the spacecraft itself. But he soon determined that only if he added to the equations an invented force—a constant change in velocity (an acceleration) back toward the Sun for every second of the trip—would the location predicted for Pioneer 10's signal match the location of its actual signal.

Had Pioneer 10 encountered something unusual along its path? If so, that could explain everything. Nope. Pioneer 11 was heading out of the solar system in a whole other direction, yet it, too, required an adjustment to its predicted location. In fact, Pioneer 11's anomaly was somewhat larger than Pioneer 10's.

Faced with either revising the tenets of conventional physics or seeking ordinary explanations for the anomaly, Anderson and his JPL collaborator Slava Turyshev chose the latter. A wise first step. You don't want to invent a new law of physics to explain a mere hardware malfunction.

Because the flow of heat energy in various directions can have unexpected effects, one of the things Anderson and Turyshev looked at was the spacecraft's material self—specifically the way heat would be absorbed, conducted, and radiated from one surface to another. Their inquiry managed to account for about a tenth of the anomaly. But neither investigator is a thermal engineer. A wise second step: find one. So in early 2006 Turyshev sought out Gary Kinsella, a JPL colleague who until that moment had never met either him or a Pioneer face to face, and Turyshev convinced Kinsella to take the thermal issues to the next level. In the spring of 2007, all three men came to the Hayden Planetarium in New York City to tell a sellout crowd about their still-unfinished travails. Meanwhile, other researchers worldwide were taking up the challenge.

Consider what it's like to be a spacecraft living and working hundreds of millions of miles from the Sun. First of all, your sunny side warms up while the unheated hardware on your shady side can plunge to –455° Fahrenheit, the background temperature of outer space. Next, you're constructed of many different kinds of materials and have multiple appendages, all of which have different thermal properties and thus absorb, conduct, emit, and scatter heat differently, both within your various cavities and outside to space. In addition, your parts like to operate at very different temperatures: your cryogenic science instruments do fine in the frigidity of outer space, but your cameras favor room temperature, and your rocket thrusters, when fired, register 2,000° F. Not only that, every piece of your hardware sits within ten feet of all your other pieces of hardware.

The task facing Kinsella and his team of engineers was to assess and quantify the directional thermal influence of every feature aboard Pioneer 10. To do that, they created a computer model representing the spacecraft surrounded by a spherical envelope. Then they subdivided that surface into 2,600 zones, enabling them to track the flow of heat from every spot in the spacecraft to and through every spot in the surrounding sphere. To strengthen their case, they also hunted through all available project documents and data files, many of which hail from the days when computers relied on punch cards for data entry and stored data on nine-track tape. (Without emergency funds from the Planetary Society, by the way, those irreplaceable archives would shortly have ended up in a Dumpster.)

For the simulated world of the team's computer model, the spacecraft was placed at a test distance from the Sun (25 AU) and at a specific angle to the Sun, and all the parts were presumed to be working as they were supposed to. Kinsella and his crew determined that, indeed, the uneven thermal emission from the spacecraft's exterior surfaces does create an anomaly—and that it is indeed a continuous, sunward change in velocity.

But how much of the Pioneer anomaly can be blamed on this effect? Certainly some. Perhaps most. Possibly all.

So what about any remaining unexplained portion of the anomaly? Do we sweep it under the cosmic rug in hopes that additional Kinsellan analysis will eventually resolve the entire anomaly? Or do we carefully reconsider the accuracy and inclusiveness of Newton's laws of gravity, as a few zealous physicists have been doing for a couple of decades?

Pre-Pioneers, Newtonian gravity had never been measured—and was therefore never confirmed—with great precision over great distances. In fact, Slava Turyshev, an expert in Einstein's general relativity, regards the Pioneers as (unintentionally) the largest-ever gravitational experiment to confirm whether Newtonian gravity is fully valid in the outer solar system. That experiment, he contends, shows it might not be. In addition, as any physicist can demonstrate, beyond 15 AU the effects of Einsteinian gravity are negligible.

In early 2009, for the benefit of visitors to the Planetary Society's website, Turyshev and his colleague Viktor Toth eloquently explained why they've kept plugging away at the Pioneer anomaly. Their explanation, titled "Finding a needle in the haystack or proving that there may be none," is worth quoting at length:

> In the short run, knowing the gravitational constant to one more decimal digit of precision or placing even tighter limits on any deviation from Einstein's gravitational theory may seem like painfully nitpicking detail. Yet one must not lose sight of the "big picture." When researchers were measuring the properties of electricity with ever more refined instruments over two hundred years ago, they did not envision continent-spanning power grids, an information economy, or tiny electrical signals reaching us from the unfathomable depths of the outer solar system, sent by man-made machines. They just performed meticulous experiments laying down the laws connecting electricity to magnetism or the electromotive force to chemical reactions. Yet their work paved the way to our modern society.
>
> Similarly, we cannot envision today what research into gravitational science will bring tomorrow. Perhaps one day humankind will harness gravity. Perhaps one day a trip across the solar system using a yet-to-be-devised gravity engine may not seem a bigger deal than crossing an ocean in a jetliner today. Perhaps one day human beings will travel to the stars in spacecraft that no longer need rockets. Who knows? But one thing we know for sure: none of that will happen unless we do a meticulous job today. Our work, whether it proves the existence of gravity beyond Einstein or just improves the navigation of spacecraft in deep space by accounting for a small thermal recoil force with precision, lays down the foundations that may, one day, lead to such dreams.

For the time being, though, two forces seem to be at play in deep space: Newton's laws of gravity and the mysterious Pioneer anomaly. Until the anomaly is thoroughly accounted for by misbehaving hardware, and can therefore be eliminated from consideration, Newton's laws will remain unconfirmed. And there just might be a rug somewhere in the cosmos with a new law of physics under it, waiting to be uncovered.

... CHAPTER THIRTY-SIX

WHAT NASA MEANS TO AMERICA'S FUTURE*

I wish I had a nickel for every time someone said, "Why are we spending money up there when we have problems down here?" The first and simplest answer to that concern is that one day there'll be a killer asteroid headed straight for us, which means not all your problems are Earth-based. At some point, you've also got to look up.

Under President Barack Obama's space plan, NASA will be promoting commercial access to low Earth orbit. The National Aeronautics and Space Act of 1958 makes NASA responsible for advancing the space frontier. And since low Earth orbit is no longer a space frontier, NASA must move to the next step. The current plan says we're not going to the Moon anymore and recommends we go to Mars one day—I don't know when.

I'm worried by this scenario. Without an actual plan to go somewhere beyond low Earth orbit, we've got nothing to shape the career dreams of young America. As best as I can judge, NASA is like a force of nature unto itself, capable of stimulating the formation of scientists, engineers, mathematicians, and technologists—the STEM research fields. You nurture these people for the sake of society, and they become the ones who make tomorrow happen.

The strength of economies in the twenty-first century will derive from the investments made in science and technology. This is something we've

* Adapted from Q&A, University of Buffalo Distinguished Speaker Series, March 31, 2010.

witnessed since the dawn of the Industrial Revolution: the nations that have embraced those investments are the nations that have led the world.

America is fading right now. Nobody's dreaming about tomorrow anymore. NASA knows how to dream about tomorrow—if the funding can accommodate it, if the funding can empower it, if the funding can enable it. Sure, you need good teachers. But the teachers come and go, because kids go on to the next grade and then the grade after that. Teachers can help light a flame, but we need something to keep the flame fanned. And that's the effect of NASA on who and what we are as a nation, what we have been as a nation, and perhaps for a while took for granted as a nation. Today the most powerful particle accelerator in the world is hundreds of feet underground at the border between France and Switzerland. The world's fastest train is made by Germans and is running in China. Meanwhile, here in America I see our infrastructure collapsing and no one dreaming about tomorrow.

Everybody thinks they can put a Band-Aid on this or that problem. Meanwhile, the agency with the most power to shape the dreams of a nation is currently underfunded to do what it must be doing—which is to make those dreams come true. And doing it for half a penny on a dollar.

How much would *you* pay for the universe?

Space Tweet #67

The US military spends as much in 23 days as NASA spends in a year – and that's when we're not fighting a war

Jul 8, 2011 11:13 AM

EPILOGUE

The Cosmic Perspective*

> Of all the sciences cultivated by mankind, Astronomy is acknowledged to be, and undoubtedly is, the most sublime, the most interesting, and the most useful. For, by knowledge derived from this science, not only the bulk of the Earth is discovered . . . ; but our very faculties are enlarged with the grandeur of the ideas it conveys, our minds exalted above [their] low contracted prejudices.
>
> —JAMES FERGUSON, *Astronomy Explained Upon Sir Isaac Newton's Principles, And Made Easy To Those Who Have Not Studied Mathematics* (1757)

Long before anyone knew that the universe had a beginning, before we knew that the nearest large galaxy lies more than two million light-years from Earth, before we knew how stars work or whether atoms exist, James Ferguson's enthusiastic introduction to his favorite science rang true. Yet his words, apart from their eighteenth-century flourish, could have been written yesterday.

But who gets to think that way? Who gets to celebrate this cosmic view of life? Not the migrant farmworker. Not the sweatshop worker. Certainly not the homeless person rummaging through the trash for food. You need the luxury of time not spent on mere survival. You need to live in a nation whose government values the search to understand humanity's place in the universe. You need a society in which intellectual pursuit can take you to the frontiers of discovery, and in which news of your discoveries can be

* Adapted from "The Cosmic Perspective," *Natural History,* April 2007.

routinely disseminated. By those measures, most citizens of industrialized nations do quite well.

Yet the cosmic view comes with a hidden cost. When I travel thousands of miles to spend a few moments in the fast-moving shadow of the Moon during a total solar eclipse, sometimes I lose sight of Earth.

When I pause and reflect on our expanding universe, with its galaxies hurtling away from one another, embedded within the ever-stretching, four-dimensional fabric of space and time, sometimes I forget that uncounted people walk this Earth without food or shelter, and that children are disproportionately represented among them.

When I pore over the data that establish the mysterious presence of dark matter and dark energy throughout the universe, sometimes I forget that every day—every twenty-four-hour rotation of Earth—people kill and get killed in the name of someone else's conception of God, and that some people who do not kill in the name of God kill in the name of their nation's needs or wants.

When I track the orbits of asteroids, comets, and planets, each one a pirouetting dancer in a cosmic ballet choreographed by the forces of gravity, sometimes I forget that too many people act in wanton disregard for the delicate interplay of Earth's atmosphere, oceans, and land, with consequences that our children and our children's children will witness and pay for with their health and well-being.

And sometimes I forget that powerful people rarely do all they can to help those who cannot help themselves.

I occasionally forget those things because, however big the world is—in our hearts, our minds, and our outsize atlases—the universe is even bigger. A depressing thought to some, but a liberating thought to me.

Consider an adult who tends to the traumas of a child: a broken toy, a scraped knee, a schoolyard bully. Adults know that kids have no clue what constitutes a genuine problem, because inexperience greatly limits their childhood perspective.

As grown-ups, dare we admit to ourselves that we, too, have a collective immaturity of view? Dare we admit that our thoughts and behaviors spring from a belief that the world revolves around us? Apparently not. Yet the evidence abounds. Part the curtains of society's racial, ethnic, religious, national, and cultural conflicts, and you find the human ego turning the knobs and pulling the levers.

Now imagine a world in which everyone, but especially people with power and influence, holds an expanded view of our place in the cosmos. With that perspective, our problems would shrink—or never arise at all—and we could celebrate our earthly differences while shunning the behavior of our predecessors who slaughtered each other because of them.

Back in February 2000, the newly rebuilt Hayden Planetarium featured a space show called "Passport to the Universe," written by Ann Druyan and Steven Soter (collaborators with Carl Sagan on the original *Cosmos* TV series). The show took visitors on a virtual zoom from New York City to the deepest regions of space. En route the audience saw Earth, then the solar system, then the Milky Way galaxy's hundreds of billions of stars shrink to barely visible dots on the planetarium's dome.

Within a month of opening day, I received a letter from an Ivy League professor of psychology whose expertise was things that make people feel insignificant. I never knew one could specialize in such a field. He wanted to administer a before-and-after questionnaire to visitors, assessing the depth of their depression after viewing "Passport to the Universe." The show, he wrote, elicited the most dramatic feelings of smallness he had ever experienced.

How could that be? Every time I see the space show (and others we've produced), I feel alive and spirited and connected. I also feel large, knowing that the goings-on within the three-pound human brain are what enabled us to figure out our place.

Allow me to suggest that it's the professor, not I, who has misread nature. His ego was too big to begin with, inflated by delusions of significance and fed by cultural assumptions that human beings are more important than everything else.

In all fairness to the fellow, powerful forces in society leave most of us susceptible. As was I . . . until the day I learned in biology class that more bacteria live and work in one centimeter of my colon than the number of people who have ever existed in the world. That kind of information makes you think twice about who—or what—is actually in charge.

From that day on, I began to think of people not as the masters of space and time but as participants in a great cosmic chain of being, with a direct genetic link across species both living and extinct, extending back nearly four billion years to the earliest single-celled organisms on Earth.

I know what you're thinking: we're smarter than bacteria.

No doubt about it, we're smarter than every other living creature that ever walked, crawled, or slithered on Earth. But how smart is that? We cook our food. We compose poetry and music. We do art and science. We're good at math. Even if you're bad at math, you're probably much better at it than the smartest chimpanzee, whose genetic identity varies in only trifling ways from ours. Try as they might, primatologists will never get a chimpanzee to learn the multiplication table or do long division.

If small genetic differences between us and our fellow apes account for our vast difference in intelligence, maybe that difference in intelligence is not so vast after all.

Imagine a life-form whose brainpower is to ours as ours is to a chimpanzee's. To such a species, our highest mental achievements would be trivial. Their toddlers, instead of learning their ABCs on *Sesame Street,* would learn multivariable calculus on *Boolean Boulevard.* Our most complex theorems, our deepest philosophies, the cherished works of our most creative artists, would be projects their schoolkids bring home for Mom and Dad to display on the refrigerator door. These creatures would study Stephen Hawking (who occupies the same endowed professorship once held by Newton at the University of Cambridge) because he's slightly more clever than other humans, owing to his ability to do theoretical astrophysics and other rudimentary calculations in his head.

If a huge genetic gap separated us from our closest relative in the animal kingdom, we could justifiably celebrate our brilliance. We might be entitled to walk around thinking we're distant and distinct from our fellow creatures. But no such gap exists. Instead, we are one with the rest of nature, fitting neither above nor below, but within.

Need more ego softeners? Simple comparisons of quantity, size, and scale do the job well.

Take water. It's simple, common, and vital. There are more molecules of water in an eight-ounce cup of the stuff than there are cups of water in all the world's oceans. Every cup that passes through a single person and eventually rejoins the world's water supply holds enough molecules to mix fifteen hundred of them into every other cup of water in the world. No way around it: some of the water you just drank passed through the kidneys of Socrates, Genghis Khan, and Joan of Arc.

How about air? Also vital. A single breathful draws in more air molecules than there are breathfuls of air in Earth's entire atmosphere. That means some of the air you just breathed passed through the lungs of Napoleon, Beethoven, Lincoln, and Billy the Kid.

Time to get cosmic. There are more stars in the universe than grains of sand on any beach, more stars than seconds have passed since Earth formed, more stars than words and sounds ever uttered by all the humans who ever lived.

Want a sweeping view of the past? Our unfolding cosmic perspective takes you there. Light takes time to reach Earth's observatories from the depths of space, and so you see objects and phenomena not as they are but as they once were. That means the universe acts like a giant time machine: the farther away you look, the further back in time you see—back almost to the beginning of time itself. Within that horizon of reckoning, cosmic evolution unfolds continuously, in full view.

Want to know what we're made of? Again, the cosmic perspective offers a bigger answer than you might expect. The chemical elements of the universe are forged in the fires of high-mass stars that end their lives in stupendous explosions, enriching their host galaxies with the chemical arsenal of life as we know it. The result? The four most common chemically active elements in the universe—hydrogen, oxygen, carbon, and nitrogen—are the four most common elements of life on Earth. We are not simply in the universe. The universe is in us.

Yes, we are stardust. But we may not be of this Earth. Several separate lines of research, when considered together, have forced investigators to reassess who we think we are and where we think we came from.

First, computer simulations show that when a large asteroid strikes a planet, the surrounding areas can recoil from the impact energy, catapulting rocks into space. From there, they can travel to—and land on—other planetary surfaces. Second, microorganisms can be hardy. Some survive the extremes of temperature, pressure, and radiation inherent in space travel. If the rocky flotsam from an impact hails from a planet with life, microscopic fauna could have stowed away in the rocks' nooks and crannies. Third, recent evidence suggests that shortly after the formation of our solar system, Mars was wet, and perhaps fertile, even before Earth was.

Those findings mean it's conceivable that life began on Mars and later seeded life on Earth, a process known as panspermia. So all earthlings might—just might—be descendants of Martians.

Again and again across the centuries, cosmic discoveries have demoted our self-image. Earth was once assumed to be astronomically unique, until astronomers learned that Earth is just another planet orbiting the Sun. Then we presumed the Sun was unique, until we learned that the countless stars of the night sky are suns themselves. Then we presumed our galaxy, the Milky Way, was the entire known universe, until we established that the countless fuzzy things in the sky are other galaxies, dotting the landscape of our known universe.

Today, how easy it is to presume that one universe is all there is. Yet emerging theories of modern cosmology, as well as the continually reaffirmed improbability that anything is unique, require that we remain open to the latest assault on our plea for distinctiveness: multiple universes, otherwise known as the multiverse, in which ours is just one of countless bubbles bursting forth from the fabric of the cosmos.

The cosmic perspective flows from fundamental knowledge. But it's more than just what you know. It's also about having the wisdom and insight to apply that knowledge to assessing our place in the universe. And its attributes are clear:

The cosmic perspective comes from the frontiers of science, yet it is not solely the provenance of the scientist. It belongs to everyone.

The cosmic perspective is humble.

The cosmic perspective is spiritual—even redemptive—but not religious.

The cosmic perspective enables us to grasp, in the same thought, the large and the small.

The cosmic perspective opens our minds to extraordinary ideas but does not leave them so open that our brains spill out, making us susceptible to believing anything we're told.

The cosmic perspective opens our eyes to the universe, not as a benevolent cradle designed to nurture life but as a cold, lonely, hazardous place.

The cosmic perspective shows Earth to be a mote, but a precious mote and, for the moment, the only home we have.

The cosmic perspective finds beauty in the images of planets, moons, stars, and nebulae but also celebrates the laws of physics that shape them.

The cosmic perspective enables us to see beyond our circumstances, allowing us to transcend the primal search for food, shelter, and sex.

The cosmic perspective reminds us that in space, where there is no air, a flag will not wave—an indication that perhaps flag waving and space exploration do not mix.

The cosmic perspective not only embraces our genetic kinship with all life on Earth but also values our chemical kinship with any yet-to-be discovered life in the universe, as well as our atomic kinship with the universe itself.

At least once a week, if not once a day, we might each ponder what cosmic truths lie undiscovered before us, perhaps awaiting the arrival of a clever thinker, an ingenious experiment, or an innovative space mission to reveal them. We might further ponder how those discoveries may one day transform life on Earth.

Absent such curiosity, we are no different from the provincial farmer who expresses no need to venture beyond the county line, because his forty acres meet all his needs. Yet if all our predecessors had felt that way, the farmer would instead be a cave dweller, chasing down his dinner with a stick and a rock.

During our brief stay on planet Earth, we owe ourselves and our descendants the opportunity to explore—in part because it's fun to do. But there's a far nobler reason. The day our knowledge of the cosmos ceases to expand, we risk regressing to the childish view that the universe figuratively and literally revolves around us. In that bleak world, arms-bearing, resource-hungry people and nations would be prone to act on their "low contracted prejudices." And that would be the last gasp of human enlightenment—until the rise of a visionary new culture that could once again embrace the cosmic perspective.

APPENDIX A

National Aeronautics and Space Act of 1958, As Amended

Source: National Aeronautics and Space Administration.

NATIONAL AERONAUTICS AND SPACE ACT OF 1958

Pub. L. No. 85-568

72 Stat. 426-438 (Jul. 29, 1958)

As Amended

AN ACT

To provide for research into problems of flight within and outside the earth's atmosphere, and for other purposes.

Be it enacted by the Senate and House of Representatives of the United States of America in Congress assembled,

TITLE I—SHORT TITLE, DECLARATION OF POLICY, AND DEFINITIONS

SHORT TITLE

Sec. 101. This Act may be cited as the "National Aeronautics and Space Act of 1958"

DECLARATION OF POLICY AND PURPOSE

Sec. 102. (a) The Congress hereby declares that it is the policy of the United States that activities in space should be devoted to peaceful purposes for the benefit of all mankind.

(b) The Congress declares that the general welfare and security of the United States require that adequate provision be made for aeronautical and space activities. The Congress further declares that such activities shall be the responsibility of, and shall be directed by, a civilian agency exercising control over aeronautical and space activities sponsored by the United States, except that activities peculiar to or primarily associated with the development of weapons systems, military operations, or the defense of the United States (including the research and development necessary to make effective provision for the defense of the United States) shall be the responsibility of, and shall be directed by, the Department of Defense; and that determination as to which such agency has responsibility for and direction of any such activity shall be made by the President in conformity with section 2471(e).

(c) The Congress declares that the general welfare of the United States requires that the National Aeronautics and Space Administration (as established by title II of this Act) seek and encourage, to the maximum extent possible, the fullest commercial use of space.

(d) The aeronautical and space activities of the United States shall be conducted so as to contribute materially to one or more of the following objectives:

(1) The expansion of human knowledge of the Earth and of phenomena in the atmosphere and space;

(2) The improvement of the usefulness, performance, speed, safety, and efficiency of aeronautical and space vehicles;

(3) The development and operation of vehicles capable of carrying instruments, equipment, supplies, and living organisms through space;

(4) The establishment of long-range studies of the potential benefits to be gained from, the Opportunities for, and the problems involved in the utilization of aeronautical and space activities for peaceful and scientific purposes;

(5) The preservation of the role of the United States as a leader in aeronautical and space science and technology and in the application thereof to the conduct of peaceful activities within and outside the atmosphere;

(6) The making available to agencies directly concerned with national defense of discoveries that have military value or significance, and the furnishing by such agencies, to the civilian agency established to direct and control nonmilitary aeronautical and space activities, of information as to discoveries which have value or significance to that agency;

(7) Cooperation by the United States with other nations and groups of nations in work done pursuant to this Act and in the peaceful application of the results thereof;

(8) The most effective utilization of the scientific and engineering resources of the United States, with close cooperation among all interested agencies of the United States in order to avoid unnecessary duplication of effort, facilities, and equipment; and

(9) The preservation of the United States preeminent position in aeronautics and space through research and technology development related to associated manufacturing processes.

(e) The Congress declares that the general welfare of the United States requires that the unique competence in scientific and engineering systems of the National Aeronautics and Space Administration also be directed toward ground propulsion systems research and development. Such development shall be conducted so as to contribute to the objectives of developing energy- and petroleum-conserving ground propulsion systems, and of minimizing the environmental degradation caused by such systems.

(f) The Congress declares that the general welfare of the United States requires that the unique competence of the National Aeronautics and Space Administration in science and engineering systems be directed to assisting in bioengineering research, development, and demonstration programs designed to alleviate and minimize the effects of disability.

(g) The Congress declares that the general welfare and security of the United States require that the unique competence of the National Aeronautics and Space

Administration be directed to detecting, tracking, cataloguing, and characterizing near-Earth asteroids and comets in order to provide warning and mitigation of the potential hazard of such near-Earth objects to the Earth.

(h) It is the purpose of this Act to carry out and effectuate the policies declared in subsections (a), (b), (c), (d), (e), (f), and (g).

DEFINITIONS

Sec. 103. As used in this Act-

(1) the term "aeronautical and space activities" means (A) research into, and the solution of, problems of flight within and outside the Earth's atmosphere, (B) the development, construction, testing, and operation for research purposes of aeronautical and space vehicles, (C) the operation of a space transportation system including the Space Shuttle, upper stages, space platforms, and related equipment, and (D) such other activities as may be required for the exploration of space; and

(2) the term "aeronautical and space vehicles" means aircraft, missiles, satellites, and other space vehicles, manned and unmanned, together with related equipment, devices, components, and parts.

TITLE II—COORDINATION OF AERONAUTICAL AND SPACE ACTIVITIES

NATIONAL AERONAUTICS AND SPACE COUNCIL

Sec. 201. (a) [There is hereby established the National Aeronautics and Space Council. . . .] abolished.

NATIONAL AERONAUTICS AND SPACE ADMINISTRATION

Sec. 202. (a) There is hereby established the National Aeronautics and Space Administration (hereinafter called the "Administration"). The Administration shall be headed by an Administrator, who shall be appointed from civilian life by the President, by and with the advice and consent of the Senate. Under the supervision and direction of the President, the Administrator shall be responsible for the exercise of all powers and the discharge of all duties of the Administration, and shall have authority and control over all personnel and activities thereof.

(b) There shall be in the Administration a Deputy Administrator, who shall be appointed from civilian life by the President by and with the advice and consent of the Senate and shall perform such duties and exercise such powers as the Administrator may prescribe. The Deputy Administrator shall act for, and exercise the powers of, the Administrator during his absence or disability.

(c) The Administrator and the Deputy Administrator shall not engage in any other business, vocation, or employment while serving as such.

FUNCTIONS OF THE ADMINISTRATION

Sec. 203. (a) The Administration, in order to carry out the purpose of this Act, shall–

(1) plan, direct, and conduct aeronautical and space activities;

(2) arrange for participation by the scientific community in planning scientific measurements and observations to be made through use of aeronautical and space vehicles, and conduct or arrange for the conduct of such measurements and observations;

(3) provide for the widest practicable and appropriate dissemination of information concerning its activities and the results thereof;

(4) seek and encourage, to the maximum extent possible, the fullest commercial use of space; and

(5) encourage and provide for Federal Government use of commercially provided space services and hardware, consistent with the requirements of the Federal Government.

(b) (1) The Administration shall, to the extent of appropriated funds, initiate, support, and carry out such research, development, demonstration, and other related activities in ground propulsion technologies as are provided for in sections 4 through 10 of the Electric and Hybrid Vehicle Research, Development, and Demonstration Act of 1976.

(2) The Administration shall initiate, support, and carry out such research, development, demonstrations, and other related activities in solar heating and cooling technologies (to the extent that funds are appropriated therefor) as are provided for in sections 5, 6, and 9 of the Solar Heating and Cooling Demonstration Act of 1974.

(c) In the performance of its functions the Administration is authorized

(1) to make, promulgate, issue, rescind, and amend rules and regulations governing the manner of its operations and the exercise of the powers vested in it by law;

(2) to appoint and fix the compensation of such officers and employees as may be necessary to carry out such functions. Such officers and employees shall be appointed in accordance with the civil-service laws and their compensation fixed in accordance with the Classification Act of 1949, except that (A) to the extent the Administrator deems such action necessary to the discharge of his responsibilities, he may appoint not more than four hundred and twenty-five of the scientific, engineering, and administrative personnel of the Administration without regard to such laws, and may fix the compensation of such personnel not in excess of the rate of basic pay payable for level III of the Executive Schedule, and (B) to the extent the Administrator deems such action necessary to recruit specially qualified scientific and engineering talent, he may establish the entrance grade for scientific and engineering personnel without previous service in the Federal Government at a level up to two grades higher than the grade provided for such personnel under the General

Schedule established by the Classification Act of 1949, and fix their compensation accordingly;

(3) to acquire (by purchase, lease, condemnation, or otherwise), construct, improve, repair, operate, and maintain laboratories, research and testing sites and facilities, aeronautical and space vehicles, quarters and related accommodations for employees and dependents of employees of the Administration, and such other real and personal property (including patents), or any interest therein, as the Administration deems necessary within and outside the continental United States; to acquire by lease or otherwise, through the Administrator of General Services, buildings or parts of buildings in the District of Columbia for the use of the Administration for a period not to exceed ten years without regard to the Act of March 3, 1877 (40 U.S.C. 34); to lease to others such real and personal property; to sell and otherwise dispose of real and personal property (including patents and rights thereunder) in accordance with the provisions of the Federal Property and Administrative Services Act of 1949, as amended (40 U.S.C. 471 et seq.); and to provide by contract or otherwise for cafeterias and other necessary facilities for the welfare of employees of the Administration at its installations and purchase and maintain equipment therefor;

(4) to accept unconditional gifts or donations of services, money, or property, real, personal, or mixed, tangible or intangible;

(5) without regard to section 3648 of the Revised Statutes, as amended (31 U.S.C. 529), to enter into and perform such contracts, leases, cooperative agreements, or other transactions as may be necessary in the conduct of its work and on such terms as it may deem appropriate, with any agency or instrumentality of the United States, or with any State, Territory, or possession, or with any political subdivision thereof, or with any person, firm, association, corporation, or educational institution. To the maximum extent practicable and consistent with the accomplishment of the purposes of this Act, such contracts, leases, agreements, and other transactions shall be allocated by the Administrator in a manner which will enable small-business concerns to participate equitably and proportionately in the conduct of the work of the Administration;

(6) to use, with their consent, the services, equipment, personnel, and facilities of Federal and other agencies with or without reimbursement, and on a similar basis to cooperate with other public and private agencies and instrumentalities in the use of services, equipment, and facilities. Each department and agency of the Federal Government shall cooperate fully with the Administration in making its services, equipment, personnel, and facilities available to the Administration, and any such department or agency is authorized, notwithstanding any other provision of law, to transfer to or to receive from the Administration, without reimbursement, aeronautical and space vehicles, and supplies and equipment other than administrative supplies or equipment;

(7) to appoint such advisory committees as may be appropriate for purposes of consultation and advice to the Administration in the performance of its functions;

(8) to establish within the Administration such offices and procedures as may be appropriate to provide for the greatest possible coordination of its activities under this Act with related scientific and other activities being carried on by other public and private agencies and organizations;

(9) to obtain services as authorized by section 3109 of title 5, United States Code, but at rates for individuals nor to exceed the per diem rate equivalent to the rate for GS-18;

(10) when determined by the Administrator to be necessary, and subject to such security investigations as he may determine to be appropriate, to employ aliens without regard to statutory provisions prohibiting payment of compensation to aliens;

(11) to provide by concession, without regard to section 321 of the Act of June 30, 1932 (47 Stat. 412; 40 U.S.C. 303b), on such terms as the Administrator may deem to be appropriate and to be necessary to protect the concessioner against loss of his investment in property (but not anticipated profits) resulting from the Administration's discretionary acts and decisions, for the construction, maintenance, and operation of all manner of facilities and equipment for visitors to the several installations of the Administration and, in connection therewith, to provide services incident to the dissemination of information concerning its activities to such visitors, without charge or with a reasonable charge therefor (with this authority being in addition to any other authority which the Administration may have to provide facilities, equipment, and services for visitors to its installations). A concession agreement under this paragraph may be negotiated with any qualified proposer following due consideration of all proposals received after reasonable public notice of the intention to contract. The concessioner shall be afforded a reasonable opportunity to make a profit commensurate with the capital invested and the obligations assumed, and the consideration paid by him for the concession shall be based on the probable value of such opportunity and not on maximizing revenue to the United States. Each concession agreement shall specify the manner in which the concessioner's records are to be maintained, and shall provide for access to any such records by the Administration and the Comptroller General of the United States for a period of five years after the close of the business year to which such records relate. A concessioner may be accorded a possessory interest, consisting of all incidents of ownership except legal title (which shall vest in the United States), in any structure[,] fixture, or improvement he constructs or locates upon land owned by the United States; and, with the approval of the Administration, such possessory interest may be assigned, transferred, encumbered, or relinquished by him, and, unless otherwise provided by contract, shall not be extinguished by the expiration or other termination of the concession and may not be taken for public use without just compensation;

(12) with the approval of the President, to enter into cooperative agreements under which members of the Army, Navy, Air Force, and Marine Corps may be detailed by the appropriate Secretary for services in the performance of functions under this Act to the same extent as that to which they might be lawfully assigned in the Department of Defense;

(13) (A) to consider, ascertain, adjust, determine, settle, and pay, on behalf of the United States, in full satisfaction thereof, any claim for $25,000 or less against the United States for bodily injury, death, or damage to or loss of real or personal property resulting from the conduct of the Administration's functions as specified in subsection (a) of this section, where such claim is presented to the Administration in writing within two years after the accident or incident out of which the claim arises; and

(B) if the Administration considers that a claim in excess of $25,000 is meritorious and would otherwise be covered by this paragraph, to report the facts and circumstances thereof to the Congress for its consideration. and

(14) Repealed.

CIVILIAN-MILITARY LIAISON COMMITTEE

Sec. 204. [Civilian-Military Liaison Committee] abolished.

INTERNATIONAL COOPERATION

Sec. 205. The Administration, under the foreign policy guidance of the President, may engage in a program of international cooperation in work done pursuant to this Act, and in the peaceful application of the results thereof, pursuant to agreements made by the President with the advice and consent of the Senate.

REPORTS TO CONGRESS

Sec. 206. (a) The President shall transmit to the Congress in May of each year a report, which shall include (1) a comprehensive description of the programmed activities and the accomplishments of all agencies of the United States in the field of aeronautics and space activities during the preceding fiscal year, and (2) an evaluation of such activities and accomplishments in terms of the attainment of, or the failure to attain, the objectives described in section 102(c) of this Act.

(b) Any report made under this section shall contain such recommendations for additional legislation as the Administrator or the President may consider necessary or desirable for the attainment of the objectives described in section 102(c) of this Act.

(c) No information which has been classified for reasons of national security shall be included in any report made under this section, unless such information has been declassified by, or pursuant to authorization given by, the President.

DISPOSAL OF EXCESS LAND

Sec. 207. Notwithstanding the provisions of this or any other law, the Administration may not report to a disposal agency as excess to the needs of the Administration any land having an estimated value in excess of $50,000 which is owned by the United States and under the jurisdiction and control of the Administration, unless (A) a period of thirty days has passed after the receipt by the Speaker and the Committee on Science and Astronautics of the House of Representatives and the President and the Committee on Aeronautical and Space Sciences of the Senate of a report by the Administrator or his designee containing a full and complete statement of the action proposed to be taken and the facts and circumstances relied upon in support of such action, or (B) each such committee before the expiration of such period has transmitted to the Administrator written notice to the effect that such committee has no objection to the proposed action.

DONATIONS FOR SPACE SHUTTLE ORBITER

Sec. 208. [Donations for Space Shuttle Orbiter] authority expired.

TITLE III—MISCELLANEOUS

NATIONAL ADVISORY COMMITTEE FOR AERONAUTICS

Sec. 301. (a) The National Advisory Committee for Aeronautics, on the effective date of this section, shall cease to exist. On such date all functions, powers, duties, and obligations, and all real and personal property, personnel (other than members of the Committee), funds, and records of that organization, shall be transferred to the Administration.

(b) Section 2302 of title 10 of the United States Code is amended by striking out "or the Executive Secretary of the National Advisory Committee for Aeronautics." and inserting in lieu thereof "or the Administrator of the National Aeronautics and Space Administration."; and section 2303 of such title 10 is amended by striking out "The National Advisory Committee for Aeronautics." and inserting in lieu thereof "The National Aeronautics and Space Administration."

(c) The first section of the Act of August 26, 1950 (5 U.S.C. 22-1), is amended by striking out "the Director, National Advisory Committee for Aeronautics" and inserting in lieu thereof "the Administrator of the National Aeronautics and Space Administration", and by striking out "or National Advisory Committee for Aeronautics" and inserting in lieu thereof "or National Aeronautics and Space Administrator".

(d) The Unitary Wind Tunnel Plan Act of 1949 (50 U.S.C. 511-515) is amended (1) by striking out "The National Advisory Committee for Aeronautics (hereinafter referred to as the 'Committee')" and inserting in lieu thereof "The Administrator of the National Aeronautics and Space Administration (hereinafter referred to as the 'Administrator')"; (2) by striking out "Committee" or "Committee's" wherever they appear and inserting in lieu thereof "Administrator" and "Administrator's", respectively; and (3) by striking out "its" wherever it appears and inserting in lieu thereof "his".

(e) This section shall take effect ninety days after the date of the enactment of this Act, or on any earlier date on which the Administrator shall determine, and announce by proclamation published in the Federal Register, that the Administration has been organized and is prepared to discharge the duties and exercise the powers conferred upon it by this Act.

TRANSFER OF RELATED FUNCTIONS

Sec. 302. (a) Subject to the provisions of this section, the President, for a period of four years after the date of enactment of this Act, may transfer to the Administration any functions (including powers, duties, activities, facilities, and parts of functions) of any other department or agency of the United States or of any officer or organizational entity thereof, which relate primarily to the functions, powers, and duties of the Administration as prescribed by section 203 of this Act. In connection with any such transfer, the President may, under this section or other applicable authority, provide for appropriate transfers of records, property, civilian personnel, and funds.

(b) Whenever any such transfer is made before January 1, 1959, the President shall transmit to the Speaker of the House of Representatives and the President pro tempore of the Senate a full and complete report concerning the nature and effect of such transfer.

(c) After December 31, 1958, no transfer shall be made under this section until (1) a full and complete report concerning the nature and effect of such proposed transfer has been transmitted by the President to the Congress, and (2) the first period of sixty calendar days of regular session of the Congress following the date of receipt of such report by the Congress has expired without the adoption by the Congress of a concurrent resolution stating that the Congress does not favor such transfer.

ACCESS TO INFORMATION

Sec. 303. (a) Information obtained or developed by the Administrator in the performance of his functions under this Act shall be made available for public inspection; except (A) information authorized or required by Federal statute to be withheld, (B) information classified to protect the national security; and (C) information described in subsection (b): *Provided*, That nothing in this Act shall authorize the withholding of information by the Administrator from the duly authorized committees of the Congress.

(b) The Administrator, for a period up to 5 years after the development of information that results from activities conducted under an agreement entered into under section 203(c)(5) and (6) of this Act, and that would be a trade secret or commercial or financial information that is privileged or confidential under the meaning of section 552(b)(4) of title 5, United States Code, if the Information had been obtained from a non-Federal party participating in such an agreement, may provide appropriate protections against the dissemination of such information, including exemption from subchapter II of chapter 5 of title 5, United States Code.

SECURITY REQUIREMENTS

Sec. 304. (A) The Administrator shall establish such security requirements, restrictions, and safeguards as he deems necessary in the interest of the national security. The Administrator may arrange with the Director of the Office of Personnel Management for the conduct of such security or other personnel investigations of the Administration's officers, employees, and consultants, and its contractors and subcontractors and their officers and employees, actual or prospective, as he deems appropriate; and if any such investigation develops any data reflecting that the individual who is the subject thereof is of questionable loyalty the matter shall be referred to the Federal Bureau of Investigation for the conduct of a full field investigation, the results of which shall be furnished to the Administrator.

(b) The Atomic Energy Commission may authorize any of its employees, or employees of any contractor, prospective contractor, licensee, or prospective licensee of the Atomic Energy Commission or any other person authorized to have access to Restricted Data by the Atomic Energy Commission under subsection 145b. of the Atomic Energy Act of 1954 (42 U.S.C. 2165(b)), to permit any member, officer, or employee of the Council, or the Administrator, or any officer, employee, member of an advisory committee, contractor, subcontractor, or officer or employee of a contractor or subcontractor of the Administration, to have access to Restricted Data relating to aeronautical and space activities which is required in the performance of his duties and so certified by the Council or the Administrator, as the case may be, but only if (1) the Council or Administrator or designee thereof has determined, in accordance with the established personnel security procedures and standards of the Council or Administration, that permitting such individual to have access to such Restricted Data will not endanger the common defense and security, and (2) the Council or Administrator or designee thereof finds that the established personnel and other security procedures and standards of the Council or Administration are adequate and in reasonable conformity to the standards established by the Atomic Energy Commission under section 145 of the Atomic Energy Act of 1954 (42 U.S.C. 2165). Any individual granted access to such Restricted Data pursuant to this subsection may exchange such Data with any individual who (A) is an officer or employee of the Department of Defense, or any department or agency thereof, or a member of the armed forces, or a contractor or subcontractor of any such department, agency, or armed force, or an officer or employee of any such contractor or subcontractor, and (B) has been authorized to have access to Restricted Data under the provisions of section 143 of the Atomic Energy Act of 1954 (42 U.S.C. 2163).

(c) Chapter 37 of title 18 of the United States Code (entitled Espionage and Censorship) is amended by-

(1) adding at the end thereof the following new section:

"§799. Violation of regulations of National Aeronautics and Space Administration."

"Whoever willfully shall violate, attempt to violate, or conspire to violate any regulation or order promulgated by the Administrator of the National Aeronautics and Space Administration for the protection or security of any laboratory, station, base or other facility, or part thereof, or any aircraft, missile, spacecraft, or similar vehicle, or part thereof, or other property or equipment in the custody, of the Administration, or any real or personal property or equipment in the custody of any contractor under any contract with the Administration or any subcontractor of any such contractor, shall be fined not more than $5,000, or imprisoned not more than one year, or both."

(2) adding at the end of the sectional analysis thereof the following new item:

"§799. Violation of regulations of National Aeronautics and Space Administration."

(d) Section 1114 of tide 18 of the United States Code is amended by inserting immediately, before "while engaged in the performance of his official duties" the following: "or any officer or employee of the National Aeronautics and Space Administration directed to guard and protect property of the United States under the administration and control of the National Aeronautics and Space Administration,".

(e) The Administrator may direct such of the officers and employees of the Administration as he deems necessary in the public interest to carry firearms while in the conduct of their official duties. The Administrator may also authorize such of those employees of the contractors and subcontractors of the Administration engaged in the protection of property owned by the United States and located at facilities owned by or contracted to the United States as he deems necessary in the public interest, to carry firearms while in the conduct of their official duties.

(f) Under regulations to be prescribed by the Administrator and approved by the Attorney General of the United States, those employees of the Administration and of its contractors and subcontractors authorized to carry firearms under subsection (e) may arrest without warrant for any offense against the United States committed in their presence, or for any felony cognizable under the laws of the United States if they have reasonable grounds to believe that the person to be arrested has committed or is committing such felony. Persons granted authority to make arrests by this subsection may exercise that authority only while guarding and protecting property owned or leased by, or under the control of, the United States under the administration and control of the Administration or one of its contractors or subcontractors, at facilities owned by or contracted to the Administration.

PROPERTY RIGHTS IN INVENTIONS

Sec. 305. (a) Whenever any invention is made in the performance of any work under any contract of the Administration, and the Administrator determines that

(1) the person who made the invention was employed or assigned to perform research, development, or exploration work and the invention is related to the work he was employed or assigned to perform, or that it was within the scope of his employment duties, whether or not it was made during working hours, or with a contribution by the Government of the use of Government facilities, equipment, materials, allocated funds, information proprietary to the Government, or services of Government employees during working hours; or

(2) the person who made the invention was not employed or assigned to perform research, development, or exploration work, but the invention is nevertheless related to the contract, or to the work or duties he was employed or assigned to perform, and was made during working hours, or with a contribution from the Government of the sort referred to in clause (1), such invention shall be the exclusive property of the United States, and if such invention is patentable a patent therefor shall be issued to the United States upon application made by the Administrator, unless the Administrator waives all or any part of the rights of the United States to such invention in conformity with the provisions of subsection (f) of this section.

(b) Each contract entered into by the Administrator with any party for the performance of any work shall contain effective provisions under which such party shall furnish promptly to the Administrator a written report containing full and complete technical information concerning any invention, discovery, improvement, or innovation which may be made in the performance of any such work.

(c) No patent may be issued to any applicant other than the Administrator for any invention which appears to the Under Secretary of Commerce for Intellectual Property and Director of the United States Patent and Trademark Office (hereafter in this section referred to as the "Director") to have significant utility in the conduct of aeronautical and space activities unless the applicant files with the Director, with the application or within thirty days after request therefor by the Director, a written statement executed under oath setting forth the full facts concerning the circumstances under which such invention was made and stating the relationship (if any) of such invention to the performance of any work under any contract of the Administration. Copies of each such statement and the application to which it relates shall be transmitted forthwith by the Director to the Administrator.

(d) Upon any application as to which any such statement has been transmitted to the Administrator, the Director may, if the invention is patentable, issue a patent to the applicant unless the Administrator, within ninety days after receipt of such application and statement, requests that such patent be issued to him on behalf of the United States. If, within such time, the Administrator files such a request with the Director, the Director shall transmit notice thereof to the applicant, and shall issue such patent to the Administrator unless the applicant within thirty days after receipt of such notice requests a hearing before the Board of Patent Appeals and Interferences on the question whether the Administrator is entitled under this section to receive such patent. The Board may hear and determine, in accordance with rules and procedures established for interference cases, the question so presented, and its determination shall be subject to appeal by the applicant or by the Administrator to the United States Court of Appeals for the Federal Circuit in accordance with procedures governing appeals from decisions of the Board of Patent Appeals and Interferences in other proceedings.

(e) Whenever any patent has been issued to any applicant in conformity with subsection (d), and the Administrator thereafter has reason to believe that the statement filed by the applicant in connection therewith contained any false representation of any material fact, the Administrator within five years after the date of issuance of such patent may file with the Director a request for the transfer to the Administrator of title to such patent on the records of the Director. Notice of any such request shall be transmitted by the Director to the owner of record of such patent, and title to such patent shall be so transferred to the Administrator unless within thirty days after receipt of such notice such owner of record requests a hearing before the Board of Patent Appeals and Interferences on the question whether any such false representation was contained in such statement. Such question shall be heard and determined, and determination thereof shall be subject to review, in the manner prescribed by subsection (d) for questions arising thereunder. No request made by the Administrator under this subsection for the transfer of title to any patent, and no prosecution for the violation of any criminal statute, shall be barred by any failure of the Administrator to make a request under subsection (d) for the issuance of such patent to him, or by any notice previously given by the Administrator stating that he had no objection to the issuance of such patent to the applicant therefor.

(f) Under such regulations in conformity with this subsection as the Administrator to shall prescribe, he may waive all or any part of the rights of the United States under this section with respect to any invention or class of inventions made or which may be made by any person or class of persons in the performance of any work required by any contract of the Administration if the Administrator determines that the interests of the United States will be served thereby. Any such waiver may be made upon such terms and under such conditions as the Administrator shall determine to be required for the protection of the interests of the United States. Each such waiver made with respect to any invention shall be subject to the reservation by the Administrator of an irrevocable, nonexclusive, nontransferable, royalty-free license for the practice of such invention throughout the world by or on behalf of the United States or any foreign government pursuant to any treaty or agreement with the United States. Each proposal for any waiver under this subsection shall be referred to an Inventions and Contribution Board which shall be established by the Administrator within the Administration. Such Board shall accord to each interested party an opportunity for hearing, and shall transmit to the Administrator its findings of fact with respect to such proposal and its recommendations for action to be taken with respect thereto.

(g) [Repealed]

(h) The Administrator is authorized to take all suitable and necessary steps to protect any invention or discovery to which he has title, and to require that contractors or persons who retain title to inventions or discoveries under this section protect the inventions or discoveries to which the Administration has or may acquire a license of use.

(i) The Administration shall be considered a defense agency of the United States for the purpose of chapter 17 of title 35 of the United States Code.

(j) As used in this section

(1) the term "person" means any individual, partnership, corporation, association, institution, or other entity;

(2) the term "contract" means any actual or proposed contract, agreement, understanding, or other arrangement, and includes any assignment, substitution of parties, or subcontract executed or entered into thereunder; and

(3) the term "made", when used in relation to any invention, means the conception or first actual reduction to practice of such invention.

(k) Any object intended for launch, launched, or assembled in outer space shall be considered a vehicle for the purpose of section 272 of title 35, United States Code.

(l) The use or manufacture of any patented invention incorporated in a space vehicle launched by the United States Government for a person other than the United States shall not be considered to be a use or manufacture by or for the United States within the meaning of section 1498(a) of title 28, United States Code unless the Administration gives an express authorization or consent for such use or manufacture.

CONTRIBUTIONS AWARDS

Sec. 306. (a) Subject to the provisions of this section, the Administrator is authorized, upon his own initiative or upon application of any person, to make a monetary award, in such amount and upon such terms as he shall determine to be warranted, to any person (as defined by section 305) for any scientific or technical contribution to the Administration which is determined by the Administrator to have significant value in the conduct of aeronautical and space activities. Each application made for any such award shall be referred to the Inventions and Contributions Board established under section 305 of this Act. Such Board shall accord to each such applicant an opportunity for hearing upon such application, and shall transmit to the Administrator its recommendation as to the terms of the award, if any, to be made to such applicant for such contribution. In determining the terms and conditions of any award the Administrator shall take into account-

(1) the value of the contribution to the United States;

(2) the aggregate amount of any sums which have been expended by the applicant for the development of such contribution;

(3) the amount of any compensation (other than salary received for services rendered as an officer or employee of the Government) previously received by the applicant for or on account of the use of such contribution by the United States; and

(4) such other factors as the Administrator shall determine to be material.

(b) If more than one applicant under subsection (a) claims an interest in the same contribution, the Administrator shall ascertain and determine the respective interests of such applicants, and shall apportion any award to be made with respect to such contribution among such applicants in such proportions as he shall determine to be equitable. No award may be made under subsection (a) with respect to any contribution–

(1) unless the applicant surrenders, by such means as the Administrator shall determine to be effective, all claims which such applicant may have to receive any compensation (other than the award made under this section) for the use of such contribution or any element thereof at any time by or on behalf of the United States, or by or on behalf of any foreign government pursuant to any treaty or agreement with the United States, within the United States or at any other place;

(2) in any amount exceeding $100,000, unless the Administrator has transmitted to the appropriate committees of the Congress a full and complete report concerning the amount and terms of, and the basis for, such proposed award, and thirty calendar days of regular session of the Congress have expired after receipt of such report by such committees.

DEFENSE OF CERTAIN MALPRACTICE AND NEGLIGENCE SUITS

Sec. 307. (a) The remedy against the United States provided by sections 1346(b) and 2672 of title 28, United States Code, for damages for personal injury, including death, caused by the negligent or wrongful act or omission of any physician, dentist, nurse, pharmacist, or paramedical or other supporting personnel (including medical and dental technicians, nursing assistants, and therapists) of the Administration in the performance of medical, dental, or related health care functions (including clinical studies and investigations) while acting within the scope of his duties or employment therein or therefor shall hereafter be exclusive of any other civil action or proceeding by reason of the same subject matter against such physician, dentist, nurse, pharmacist, or paramedical or other supporting personnel (or the estate of such person) whose act or omission gave rise to such action or proceeding.

(b) The Attorney General shall defend any civil action or proceeding brought

in any court against any person referred to in subsection (a) of this section (or the estate of such person) for any such injury. Any such person against whom such civil action or proceeding is brought shall deliver within such time after date of service or knowledge of service as determined by the Attorney General, all process served upon such person or an attested true copy thereof to such person's immediate superior or to whomever was designated by the Administrator to receive such papers and such person shall promptly furnish copies of the pleading and process therein to the United States Attorney for the district embracing the place wherein the proceeding is brought to the Attorney General and to the Administrator.

(c) Upon a certification by the Attorney General that any person described in subsection (a) was acting in the scope of such person's duties or employment at the time of the incident out of which the suit arose, any such civil action or proceeding commenced in a State court shall be removed without bond at any time before trial by the Attorney General to the district court of the United States of the district and division embracing the place wherein it is pending and the proceeding deemed a tort action brought against the United States under the provisions of title 28, United States Code, and all references thereto. Should a United States district court determine on a hearing on a motion to remand held before a trial on the merits that the case so removed is one in which a remedy by suit within the meaning of subsection (a) of this section is not available against the United States, the case shall be remanded to the State court.

(d) The Attorney General may compromise or settle any claim asserted in such civil action or proceeding in the manner provided in section 2677 of title 28, United States Code, and with the same effect.

(e) For purposes of this section, the provisions of section 2680(h) of title 28, United States Code, shall not apply to any cause of action arising out of a negligent or wrongful act of omission in the performance of medical, dental, or related health care functions (including clinical studies and investigations).

(f) The Administrator or his designee may, to the extent that the Administrator or his designee deem appropriate, hold harmless or provide liability insurance for any person described in subsection (a) for damages for personal injury, including death, caused by such person's negligent or wrongful act or omission in the performance of medical, dental, or related health care functions (including clinical studies and investigations) while acting within the scope of such person's duties if such person is assigned to a foreign country or detailed for service with other than a Federal department, agency, or instrumentality or if the circumstances are such as are likely to preclude the remedies of third persons against the United States described in section 2679(b) of title 28, United States Code, for such damage or injury.

INSURANCE AND INDEMNIFICATION

Sec. 308. (a) The Administration is authorized on such terms and to the extent it may deem appropriate to provide liability insurance for any user of a space vehicle to compensate all or a portion of claims by third parties for death, bodily injury, or loss of or damage to property resulting from activities carried on in connection with the launch, operations or recovery of the space vehicle. Appropriations available to the Administration may be used to acquire such insurance, but such appropriations shall be reimbursed to the maximum extent practicable by the users under reimbursement policies established pursuant to section 203(c) of this Act.

(b) Under such regulations in conformity with this section as the Administrator shall prescribe taking into account the availability, cost and terms of liability insurance, any agreement between the Administration and a user of a space vehicle may provide that the United States will indemnify the user against claims (including reasonable expenses of litigation or settlement) by third parties for death, bodily injury, or loss of or damage to property resulting from activities carried on in connection with the launch, operations or recovery of the space vehicle, but only to the extent that such claims are not compensated by liability insurance of the user: Provided, That such indemnification may be limited to claims resulting from other than the actual negligence or willful misconduct of the user.

(c) An agreement made under subsection (b) that provides indemnification must also provide for-

(1) notice to the United States of any claim or suit against the user for the death, bodily injury, or loss of or damage to the property; and

(d) No payment may be made under subsection (b) unless the Administrator or his designee certifies that the amount is just and reasonable.

(e) Upon the approval by the Administrator, payments under subsection (b) may be made, at the Administrator's election, either from funds available for research and development not otherwise obligated or from funds appropriated for such payments.

(f) As used in this section

(1) the term "space vehicle" means an object intended for launch, launched or assembled in outer space, including the Space Shuttle and other components of a space transportation system, together with related equipment, devices, components and parts;

(2) the term "user" includes anyone who enters into an agreement with the Administration for use of all or a portion of a space vehicle, who owns or provides property to be flown on a space vehicle, or who employs a person to be flown on a space vehicle; and

(3) the term "third party" means any person who may institute a claim against a user for death, bodily injury or loss of or damage to property.

EXPERIMENTAL AEROSPACE VEHICLE

Sec. 309. (a) The Administrator may provide liability insurance for, or indemnification to, the developer of an experimental aerospace vehicle developed or used in execution of an agreement between the Administration and the developer.

(b) Terms and Conditions.

(1) Except as otherwise provided in this section, the insurance and indemnification provided by the Administration under subsection (a) to a developer shall be provided on the same terms and conditions as insurance and indemnification is provided by the Administration under section 308 of this Act to the user of a space vehicle.

(2) Insurance.

(A) A developer shall obtain liability insurance or demonstrate financial responsibility in amounts to compensate for the maximum probable loss from claims by–

(i) a third party for death, bodily injury, or property damage, or loss resulting from an activity carried out in connection with the development or use of an experimental aerospace vehicle; and

(ii) the United States Government for damage or loss to Government property resulting from such an activity.

(B) The Administrator shall determine the amount of insurance required, but, except as provided in subparagraph (C), that amount shall not be greater than the amount required under section 70112(a)(3) of title 49, United States Code, for a launch. The Administrator shall publish notice of the Administrator's determination and the applicable amount or amounts in the Federal Register within 10 days after making the determination.

(C) The Administrator may increase the dollar amounts set forth in section 70112(a) (3)(A) of title 49, United States Code, for the purpose of applying that section under this section to a developer after consultation with the Comptroller General and such experts and consultants as may be appropriate, and after publishing notice of the increase in the Federal Register not less than 180 days before the increase goes into effect. The Administrator shall make available for public inspection, not later than the date of publication of such notice, a complete record of any correspondence received by the Administration, and a transcript of any meetings in which the Administration participated, regarding the proposed increase.

(D) The Administrator may not provide liability insurance or indemnification under subsection (a) unless the developer establishes to the satisfaction of the Administrator that appropriate safety procedures and practices are being followed in the development of the experimental aerospace vehicle.

(3) Notwithstanding subsection (a), the Administrator may not indemnify a developer of an experimental aerospace vehicle under this section unless there is an agreement between the Administration and the developer described in subsection (c).

(4) If the Administrator requests additional appropriations to make payments under this section, like the payments that may be made under section 308(b) of this Act, then the request for those appropriations shall be made in accordance with the procedures established by subsections (d) and (e) of section 70113 of title 49, United States Code.

(c) Cross-Waivers.

(1) The Administrator, on behalf of the United States, and its departments, agencies, and instrumentalities, may reciprocally waive claims with a developer or cooperating party and with the related entities of that developer or cooperating party under which each party to the waiver agrees to be responsible, and agrees to ensure that its own related entities are responsible, for damage or loss to its property for which it is responsible, or for losses, resulting from any injury or death sustained by its own employees or agents, as a result of activities connected to the agreement or use of the experimental aerospace vehicle.

(2) Limitations.

(A) A reciprocal waiver under paragraph (1) may not preclude a claim by any natural person (including, but not limited to, a natural person who is an employee of the United States, the developer, the cooperating party, or their respective subcontractors) or that natural person's estate, survivors, or subrogees for injury or death, except with respect to a subrogee that is a party to the waiver or has otherwise agreed to be bound by the terms of the waiver.

(B) A reciprocal waiver under paragraph (1) may not absolve any party of liability to any natural person (including, but not limited to, a natural person who is an employee of the United States, the developer, the cooperating party, or their respective subcontractors) or such a natural person's estate, survivors, or subrogees for negligence, except with respect to a subrogee that is a party to the waiver or has otherwise agreed to be bound by the terms of the waiver.

(C) A reciprocal waiver under paragraph (1) may not be used as the basis of a claim by the Administration, or the developer or cooperating party, for indemnification against the other for damages paid to a natural person, or that natural person's estate, survivors, or subrogees, for injury or death sustained by that natural person as a result of activities connected to the agreement or use of the experimental aerospace vehicle.

(D) A reciprocal waiver under paragraph (1) may not relieve the United States, the developer, the cooperating party, or the related entities of the developer or cooperating party, of liability for damage or loss resulting from willful misconduct.

(3) Subsection (c) applies to any waiver of claims entered into by the Administration without regard to whether it was entered into before, on, or after the date of the enactment of this Act.

(d) Definitions. In this section:

(1) Cooperating Party.-The term "cooperating party" means any person who

enters into an agreement with the Administration for the performance of cooperative scientific, aeronautical, or space activities to carry out the purposes of this Act.

(2) Developer-The term "developer" means a United States person (other than a natural person) who–

(A) is a party, to an agreement with the Administration for the purpose of developing new technology for an experimental aerospace vehicle;

(B) owns or provides property to be flown or situated on that vehicle; or

(C) employs a natural person to be flown on that vehicle.

(3) Experimental Aerospace Vehicle.-The term "experimental aerospace vehicle" means an object intended to be flown in, or launched into, orbital or suborbital flight for the purpose of demonstrating technologies necessary for a reusable launch vehicle, developed under an agreement between the Administration and a developer.

(4) Related Entity.-The term "related entity" includes a contractor or subcontractor at any tier, a supplier, a grantee, and an investigator or detailee.

(e) Relationship to Other Laws.

(1) Section 308.-This section does not apply to any object, transaction, or operation to which section 308 of this Act applies.

(2) Chapter 701 of Title 49, United States Code.-The Administrator may not provide indemnification to a developer under this section for launches subject to license under section 70117(g)(1) of title 49, United States Code.

(f) Termination

(1) In General. The provisions of this section shall terminate on December 31, 2010, except that the Administrator may extend the termination date to a date not later than September 30, 2005, if the Administrator determines that such extension is in the interests of the United States.

(2) Effect of Termination on Agreement. The termination of this section shall not terminate or otherwise affect any cross-waiver agreement, insurance agreement, indemnification agreement, or other agreement entered into under this section, except as may be provided in that agreement.

APPROPRIATIONS

Sec. 310. (a) There are hereby authorized to be appropriated such sums as may be necessary to carry out this Act, except that nothing in this Act shall authorize the appropriation of any amount for (1) the acquisition or condemnation of any real property, or (2) any other item of a capital nature (such as plant or facility acquisition, construction, or expansion) which exceeds $250,000. Sums appropriated pursuant to this subsection for the construction of facilities, or for research and development activities, shall remain available until expended.

(b) Any funds appropriated for the construction of facilities may be used for emergency repairs of existing facilities when such existing facilities are made inoperative by major breakdown, accident, or other circumstances and such repairs are deemed by the Administrator to be of greater urgency than the construction of new facilities.

(c) Notwithstanding any other provision of law, the authorization of any appropriation to the Administration shall expire (unless an earlier expiration is specifically provided) at the close of the third fiscal year following the fiscal year in which the authorization was enacted, to the extent that such appropriation has not theretofore actually been made.

MISUSE OF AGENCY NAME AND INITIALS

Sec. 311. (a) No person (as defined by section 305) may (1) knowingly use the words "National Aeronautics and Space Administration" or the letters "NASA", or any combination, variation, or colorable imitation of those words or letters either alone or in combination with other words or letters, as a firm or business name in a manner reasonably calculated to convey the impression that such firm or business has some connection with, endorsement of, or authorization from, the National Aeronautics and Space Administration which does not, in fact, exist; or (2) knowingly use those words or letters or any combination, variation, or colorable imitation thereof either alone or in combination with other words or letters in connection with any product or service being offered or made available to the public in a manner reasonably calculated to convey the impression that such product or service has the authorization, support, sponsorship, or endorsement of, or the development, use, or manufacture by or on behalf of the National Aeronautics and Space Administration which does not, in fact, exist.

(b) Whenever it appears to the Attorney General that any person is engaged in an act or practice which constitutes or will constitute conduct prohibited by subsection (a), the Attorney General may initiate a civil proceeding in a district court of the United States to enjoin such act or practice.

CONTRACTS REGARDING EXPENDABLE LAUNCH VEHICLES

Sec. 312. (a) The Administrator may enter into contracts for expendable launch vehicle services that are for periods in excess of the period for which funds are otherwise available for obligation, provide for the payment for contingent liability which may accrue in excess of available appropriations in the event the Government for its convenience terminates such contracts, and provide for advance payments reasonably related to launch vehicle and related equipment, fabrication, and acquisition costs, if any such contract limits the amount of the payments that the Federal Government is allowed to make under such contract to amounts provided in advance in appropriation Acts. Such contracts may be limited to sources within the United States when the Administrator determines that such limitation is in the public interest.

(b) If funds are not available to continue any such contract, the contract shall be terminated for the convenience of the Government, and the costs of such contract shall be paid from appropriations originally available for performance of the contract, from other, unobligated appropriations currently available for the procurement of launch services, or from funds appropriated for such payments.

FULL COST APPROPRIATIONS ACCOUNT STRUCTURE

Sec. 313. (a) (1) Appropriations for the Administration for fiscal year 2007 and thereafter shall be made in three accounts, "Science, Aeronautics, and Education", "Exploration Systems and Space Operations", and an account for amounts appropriated for the necessary expenses of the Office of the Inspector General.

(2) Within the Exploration Systems and Space Operations account, no more than 10 percent of the funds for a fiscal year for Exploration Systems may be reprogrammed for Space Operations, and no more than 10 percent of the funds for a fiscal year for Space Operations may be reprogrammed for Exploration Systems. This paragraph shall not apply to reprogramming for the purposes described in subsection (b)(2).

(3) Appropriations shall remain available for two fiscal years, unless otherwise specified in law. Each account shall include the planned full costs of Administration activities.

(b) (1) To ensure the safe, timely, and successful accomplishment of Administration missions, the Administration may transfer amounts for Federal salaries and benefits; training, travel and awards; facility and related costs; information technology services; publishing services; science, engineering, fabricating and testing services; and other administrative services among accounts, as necessary.

(c) The unexpired balances of prior appropriations to the Administration for activities authorized under this Act may be transferred to the new account established for such activity in subsection (a). Balances so transferred may be merged with funds in the newly established account and thereafter may be accounted for as one fund under the same terms and conditions.

PRIZE AUTHORITY

Sec. 314. (a) In General.–The Administration may carry out a program to competitively award cash prizes to stimulate innovation in basic and applied research, technology development, and prototype demonstration that have the potential for application to the performance of the space and aeronautical activities of the Administration. The Administration may carry out a program to award prizes only in conformity with this section.

(b) Topics.–In selecting topics for prize competitions, the Administrator shall consult widely both within and outside the Federal Government, and may empanel advisory committees.

(c) Advertising.–The Administrator shall widely advertise prize competitions to encourage participation.

(d) Requirements and Registration.–For each prize competition, the Administrator shall publish a notice in the Federal Register announcing the subject of the competition, the rules for being eligible to participate in the competition, the amount of the prize, and the basis on which a winner will be selected.

(e) Eligibility.–To be eligible to win a prize under this section, an individual or entity–

(1) shall have registered to participate in the competition pursuant to any rules promulgated by the Administrator under subsection (d);

(2) shall have complied with all the requirements under this section;

(3) in the case of a private entity, shall be incorporated in and maintain a primary place of business in the United States, and in the case of an individual, whether participating singly or in a group, shall be a citizen or permanent resident of the United States; and

(4) shall not be a Federal entity or Federal employee acting within the scope of their employment.

(f) Liability.–(1) Registered participants must agree to assume any and all risks and waive claims against the Federal Government and its related entities, except in the case of willful misconduct, for any injury, death, damage, or loss of property, revenue, or profits, whether direct, indirect, or consequential, arising from their participation in a competition, whether such injury, death, damage, or loss arises through negligence or otherwise. For the purposes of this paragraph, the term 'related entity' means a contractor or subcontractor at any tier, and a supplier, user, customer, cooperating party, grantee, investigator, or detailee.

(2) Participants must obtain liability insurance or demonstrate financial responsibility, in amounts determined by the Administrator, for claims by–

(A) a third party for death, bodily injury, or property damage, or loss resulting from an activity carried out in connection with participation in a competition, with the Federal Government named as an additional insured under the registered participant's insurance policy and registered participants agreeing to indemnify the Federal Government against third party claims for damages arising from or related to competition activities; and

(B) the Federal Government for damage or loss to Government property resulting from such an activity.

(g) Judges.–For each competition, the Administration, either directly or through an agreement under subsection (h), shall assemble a panel of qualified judges to select the winner or winners of the prize competition on the basis described pursuant to subsection (d). Judges for each competition shall include individuals from outside the Administration, including from the private sector. A judge may not–

(1) have personal or financial interests in, or be an employee, officer, director, or agent of any entity that is a registered participant in a competition; or

(2) have a familial or financial relationship with an individual who is a registered participant.

(h) Administering the Competition.–The Administrator may enter into an agreement with a private, nonprofit entity to administer the prize competition, subject to the provisions of this section.

(i) Funding.–(1) Prizes under this section may consist of Federal appropriated funds and funds provided by the private sector for such cash prizes. The Administrator may accept funds from other Federal agencies for such cash prizes. The Administrator may not give any special consideration to any private sector entity in return for a donation.

(2) Notwithstanding any other provision of law, funds appropriated for prize awards under this section shall remain available until expended, and may be transferred, reprogrammed, or expended for other purposes only after the expiration of 10 fiscal years after the fiscal year for which the funds were originally appropriated. No provision in this section permits obligation or payment of funds in violation of the Anti-Deficiency Act (31 U.S.C. 1341).

(3) No prize may be announced under subsection (d) until all the funds needed to pay out the announced amount of the prize have been appropriated or committed in writing by a private source. The Administrator may increase the amount of a prize after an initial announcement is made under subsection (d) if–

(A) notice of the increase is provided in the same manner as the initial notice of the prize; and

(B) the funds needed to pay out the announced amount of the increase have been appropriated or committed in writing by a private source.

(4) No prize competition under this section may offer a prize in an amount greater than $10,000,000 unless 30 days have elapsed after written notice has been transmitted to the Committee on Science of the House of Representatives and the Committee on Commerce, Science, and Transportation of the Senate.

(5) No prize competition under this section may result in the award of more than $1,000,000 in cash prizes without the approval of the Administrator.

(j) Use of NASA Name and Insignia.–A registered participant in a competition under this section may use the Administration's name, initials, or insignia only after prior review and written approval by the Administration.

(k) Compliance With Existing Law.–The Federal Government shall not, by virtue of offering or providing a prize under this section, be responsible for compliance by registered participants in a prize competition with Federal law, including licensing, export control, and non-proliferation laws, and related regulations.

LEASE OF NON-EXCESS PROPERTY

Sec. 315. (a) In general. The Administrator may enter into a lease under this section with any person or entity (including another department or agency of the Federal Government or an entity of a State or local government) with regard to any non-excess real property and related personal property under the jurisdiction of the Administrator.

(b) Consideration.

(1) A person or entity entering into a lease under this section shall provide cash consideration for the lease at fair market value as determined by the Administrator.

(2)

(A) The Administrator may utilize amounts of cash consideration received under this subsection for a lease entered into under this section to cover the full costs to NASA in connection with the lease. These funds shall remain available until expended.

(B) Any amounts of cash consideration received under this subsection that are not utilized in accordance with subparagraph (A) shall be deposited in a capital asset account to be established by the Administrator, shall be available for capital revitalization and construction projects and improvements of real property assets and related personal property under the jurisdiction of the Administrator, and shall remain available until expended.

(C) Amounts utilized under subparagraph (B) may not be utilized for daily operating costs.

(c) Additional terms and conditions. The Administrator may require such terms and conditions in connection with a lease under this section as the Administrator considers appropriate to protect the interests of the United States.

(d) Relationship to other lease authority. The authority under this section to lease property of NASA is in addition to any other authority to lease property of NASA under law.

(e) Lease Restrictions.

(1) NASA is not authorized to lease back property under this section during the term of the out-lease or enter into other contracts with the lessee respecting the property.

(2) NASA is not authorized to enter into an out-lease under this section unless the Administrator certifies that such out-lease will not have a negative impact on NASA's mission.

(f) Sunset.—The authority to enter into leases under this section shall expire on the date that is ten years after the date of the enactment of the Commerce, Justice, Science, and Related Agencies Appropriations Act of 2008. The expiration under this subsection of authority to enter into leases under this section shall not affect the validity or term of leases or NASA's retention of proceeds from leases entered into under this section before the date of the expiration of such authority.

RETROCESSION OF JURISDICTION

Sec. 316. (a) Notwithstanding any other provision of law, the Administrator may relinquish to a State all or part of the legislative jurisdiction of the United States over lands or interests under the control of the Administrator in that State.

(b) For purposes of this section, the term 'State' means any of the several States, the District of Columbia, the Commonwealth of Puerto Rico, the United States Virgin Islands, Guam, American Samoa, the Northern Mariana Islands, and any other commonwealth, territory, or possession of the United States.

RECOVERY AND DISPOSITION AUTHORITY

Sec. 317.(a) In General.–

(1) Control of remains.–Subject to paragraphs (2) and (3), when there is an accident or mishap resulting in the death of a crewmember of a NASA human space flight vehicle, the Administrator may take control over the remains of the crewmember and order autopsies and other scientific or medical tests.

(2) Treatment.–Each crewmember shall provide the Administrator with his or her preferences regarding the treatment accorded to his or her remains and the Administrator shall, to the extent possible, respect those stated preferences.

(3) Construction.–This section shall not be construed to permit the Administrator to interfere with any Federal investigation of a mishap or accident.

(b) Definitions.–In this section:

(1) Crewmember.–The term 'crewmember' means an astronaut or other person assigned to a NASA human space flight vehicle.

(2) NASA human space flight vehicle.–The term 'NASA human space flight vehicle' means a space vehicle, as defined in section 308(f)(1), that

(A) is intended to transport 1 or more persons;

(B) is designed to operate in outer space; and

(C) is either owned by NASA, or owned by a NASA contractor or cooperating party and operated as part of a NASA mission or a joint mission with NASA.

TITLE IV—UPPER ATMOSPHERIC RESEARCH

PURPOSE AND POLICY

Sec. 401. (a) The purpose of this title is to authorize and direct the Administration to develop and carry out a comprehensive program of research, technology, and monitoring of the phenomena of the upper atmosphere so as to provide for an understanding of and to maintain the chemical and physical integrity of the Earth's upper atmosphere.

(b) The Congress declares that is the policy of the United States to undertake an immediate and appropriate research, technology, and monitoring program that will provide for understanding the physics and chemistry of the Earth's upper atmosphere.

DEFINITIONS

Sec. 402. For the purpose of this title the term "upper atmosphere" means that portion of the Earth's sensible atmosphere above the troposphere.

PROGRAM AUTHORIZED

Sec. 403. (a) In order to carry out the purposes of this title the Administration in cooperation with other Federal agencies, shall initiate and carry out a program of research, technology, monitoring, and other appropriate activities directed to understand the physics and chemistry of the upper atmosphere.

(b) In carrying out the provisions of this title the Administration shall–

(1) arrange for participation by the scientific and engineering community, of both the Nation's industrial organizations and institutions of higher education, in planning and carrying out appropriate research, in developing necessary technology and in making necessary observations and measurements;

(2) provide, by way of grant, contract, scholarships or other arrangements, to the maximum extent practicable and consistent with other laws, for the widest practicable and appropriate participation of the scientific and engineering community in the program authorized by this title; and

(3) make all results of the program authorized by this title available to the appropriate regulatory agencies and provide for the widest practicable dissemination of such results.

INTERNATIONAL COOPERATION

Sec. 404. In carrying out the provisions of this title, the Administration, subject to the direction of the President and after consultation with the Secretary of State, shall make every effort to enlist the support and cooperation of appropriate scientists and engineers of other countries and international organizations.

APPENDIX B

Selected Statutory Provisions Applicable to NASA

Source: National Aeronautics and Space Administration.

National Space Grant College and Fellowship Act
Title II, Pub. L. No. 100-147
101 Stat. 860, 869-875 (Oct. 30, 1987)
Codified at 42 U.S.C.§§ 2486-24861

SEC. 2486. CONGRESSIONAL STATEMENT OF FINDINGS

The Congress finds that–

(1) the vitality of the Nation and the quality of life of the citizens of the Nation depend increasingly on the understanding, assessment, development, and utilization of space resources;

(2) research and development of space science, space technology, and space commercialization will contribute to the quality of life, national security, and the enhancement of commerce;

(3) the understanding and development of the space frontiers require a broad commitment and an intense involvement on the part of the Federal Government in partnership with State and local governments, private industry, universities, organizations, and individuals concerned with the exploration and utilization of space;

(4) the National Aeronautics and Space Administration, through the national space grant college and fellowship program, offers the most suitable means for such commitment and involvement through the promotion of activities that will result in greater understanding, assessment, development, and utilization; and

(5) Federal support of the establishment, development, and operation of programs and projects by space grant colleges, space grant regional consortia, institutions of higher education, institutes, laboratories, and other appropriate public and private entities is the most cost-effective way to promote such activities.

SEC. 2486a. CONGRESSIONAL STATEMENT OF PURPOSE

The purposes of this title are to–

(1) increase the understanding, assessment, development, and utilization of space resources by promoting a strong educational base, responsive research and training activities, and broad and prompt dissemination of knowledge and techniques;

(2) utilize the abilities and talents of the universities of the Nation to support and contribute to the exploration and development of the resources and opportunities afforded by the space environment;

(3) encourage and support the existence of interdisciplinary and multidisciplinary programs of space research within the university community of the Nation, to engage in integrated activities of training, research and public service, to have cooperative programs with industry, and to be coordinated with the overall program of the National Aeronautics and Space Administration;

(4) encourage and support the existence of consortia, made up of university and industry members, to advance the exploration and development of space resources in

cases in which national objectives can be better fulfilled than through the programs of single universities;

(5) encourage and support Federal funding for graduate fellowships in fields related to space; and

(6) support activities in colleges and universities generally for the purpose of creating and operating a network of institutional programs that will enhance achievements resulting from efforts under this title.

SEC. 2486b. DEFINITIONS

As used in this title, the term–

(1) "Administration" means the National Aeronautics and Space Administration;

(2) "Administrator" means the Administrator of the National Aeronautics and Space Administration;

(3) "aeronautical and space activities" has the meaning given to such term in section 103(1) of the National Aeronautics and Space Act of 1958 *[42 U.S. C - 2452(1)1;*

(4) "field related to space" means any academic discipline or field of study (including the physical, natural, and biological sciences, and engineering, space technology, education, economics, sociology, communications, planning, law, international affairs, and public administration) which is concerned with or likely to improve the understanding, assessment, development, and utilization of space;

(5) "panel" means the space grant review panel established pursuant to section 210 of this title (*42 U.S. C § 2486h*);

(6) "Person" means any individual, any public or private corporation, partnership, or other association or entity (including any space grant college, space grant regional consortium, institution of higher education, institute, or laboratory), or any State, political subdivision of a State, or agency or officer of a State or political subdivision of a State;

(7) "space environment" means the environment beyond the sensible atmosphere of the Earth;

(8) "space grant college" means any public or private institution of higher education which is designated as such by the Administrator pursuant to section 208 of this title *(42 U.S. C § 2486f);*

(9) "space grant program" means any program which–

(A) is administered by any space grant college, space grant regional consortium, institution of higher education, institute, laboratory, or State or local agency; and

(B) includes two or more projects involving education and one or more of the following activities in the fields related to space–

(i) research,

(ii) training, or

(iii) advisory services;

(10) "space grant regional consortium" means any association or other alliance which is designated as such by the Administrator pursuant to section 208 of this title (*42 U.S.C. § 24860*;

(11) "space resource" means any tangible or intangible benefit, which can only be realized from–

(A) aeronautical and space activities; or

(B) advancements in any field related to space; and

(12) "State" means any State of the United States, the District of Columbia, the Commonwealth of Puerto Rico, the Virgin Islands, Guam, American Samoa, the Commonwealth of the Northern Mariana Islands, or any other territory or possession of the United States.

SEC. 2486c. NATIONAL SPACE GRANT COLLEGE AND FELLOWSHIP PROGRAM

(a) Establishment; long-range guidelines and priorities; program evaluation

The Administrator shall establish and maintain, within the Administration, a program to be known as the national space grant college and fellowship program. The national space grant college and fellowship program shall consist of the financial assistance and other activities provided for in this title. The Administrator shall establish long-range planning guidelines and priorities, and adequately evaluate the program.

(b) Functions

Within the Administration, the program shall–

(1) apply the long-range planning guidelines and the priorities established by the Administrator under subsection (a) of this section;

(2) advise the Administrator with respect to the expertise and capabilities which are available through the national space grant college and fellowship program, and make such expertise available to the Administration as directed by the Administrator;

(3) evaluate activities conducted under grants and contracts awarded pursuant to sections 206 and 207 of this title *(42 U. S. C §§ 2486d and 2486e)* to assure that the purposes set forth in section 203 *(42 U.S. C § 2486a)* of this title are implemented;

(4) encourage other Federal departments, agencies, and instrumentalities to use and take advantage of the expertise and capabilities which are available through the national space grant college and fellowship program, on a cooperative or other basis;

(5) encourage cooperation and coordination with other Federal programs concerned with the development of space resources and fields related to space;

(6) advise the Administrator on the designation of recipients supported by the national space grant college and fellowship program and, in appropriate cases, on the termination or suspension of any such designation; and

(7) encourage the formation and growth of space grant and fellowship programs.

(c) Acceptance of gifts and donations; funds from other Federal agencies; issuance of rules and regulations

To carry out the provisions of this title, the Administrator may

(1) accept conditional or unconditional gifts or donations of services, money, or property, real, personal or mixed, tangible or intangible;

(2) accept and use funds from other Federal departments, agencies, and instrumentalities to pay for fellowships, grants, contracts, and other transactions; and

(3) issue such rules and regulations as may be necessary and appropriate.

SEC. 2486d. GRANTS OR CONTRACTS

(a) Authority of Administrator; amount

The Administrator may make grants and enter into contracts or other transactions under this subsection to assist any space grant and fellowship program or project if the Administrator finds that such program or project will carry out the purposes set forth in section 203 of this title (*42 U.S. C § 2486a*). The total amount paid pursuant to any such grant or contract may equal 66 percent, or any lesser percent, of the total cost of the space grant and fellowship program or project involved, except that this limitation shall not apply in the case of grants or contracts paid for with funds accepted by the Administrator pursuant to section 205(c)(2) of this title [*42 U.S. C § 2486c(c)(2)]*.

(b) Special grants; amount; prerequisites

The Administrator may make special grants under this subsection to carry out the purposes set forth in section 203 of this title (42 U.S.C. § 2486a). The amount of any such grant may equal 100 percent, or any lesser percent, of the total cost of the project involved. No grant may be made under this subsection, unless the Administrator finds that

(1) no reasonable means is available through which the applicant can meet the matching requirement for a grant under subsection (a) of this section;

(2) the probable benefit of such project outweighs the public interest in such matching requirement; and

(3) the same or equivalent benefit cannot be obtained through the award of a contract or grant under subsection (a) of this section or section 207 of this title (42 U.S.C. § 2486e).

(c) Application

Any person may apply to the Administrator for a grant or contract under this section. Application shall be made in such form and manner, and with such content and other submissions, as the Administrator shall by regulation prescribe.

(d) Terms and conditions; limitations; leasing-, record-keeping-, audits

(1) Any grant made, or contract entered into, under this section shall be subject to the limitations and provisions set forth in paragraphs (2) and (3) of this subsec-

tion and to such other terms, conditions and requirements as the Administrator considers necessary or appropriate.

(2) No payment under any grant or contract under this section may be applied to–

(A) the purchase of any land;

(B) the purchase, construction, preservation, or repair of any building; or

(C) the purchase or construction of any launch facility or launch vehicle.

(3) Notwithstanding paragraph (2) of this subsection, the items in subparagraphs (A), (B), and (C) of such paragraph may be leased upon written approval of the Administrator.

(4) Any person who receives or utilizes any proceeds of any grant or contract under this section shall keep such records as the Administrator shall by regulation prescribe as being necessary and appropriate to facilitate effective audit and evaluation, including records which fully disclose the amount and disposition by such recipient of such proceeds, the total cost of the program or project in connection with which such proceeds were used, and the amount, if any, of such cost which was provided through other sources. Such records shall be maintained for three years after the completion of such a program or project. The Administrator and the Comptroller General of the United States, or any of their duly authorized representatives, shall have access, for the purpose of audit and evaluation, to any books, documents, papers and records of receipts which, in the opinion of the Administrator or the Comptroller General, may be related or pertinent to such grants and contracts.

SEC. 2486e. IDENTIFICATION OF SPECIFIC NATIONAL NEEDS AND PROBLEMS RELATING TO SPACE; GRANTS OR CONTRACTS WITH RESPECT TO SUCH NEEDS OR PROBLEMS, AMOUNT, APPLICATION, TERMS AND CONDITIONS

(a) The Administrator shall identify specific national needs and problems relating to space. The Administrator may make grants or enter into contracts under this section with respect to such needs or problems. The amount of any such grant or contract may equal 100 percent, or any lesser percent, of the total cost of the project involved.

(b) Any person may apply to the Administrator for a grant or contract under this section. In addition, the Administrator may invite applications with respect to specific national needs or problems identified under subsection (a) of this section. Application shall be made in such form and manner, and with such content and other submissions, as the Administrator shall by regulation prescribe. Any grant made, or contract entered into, under this section shall be subject to the limitations and provisions set forth in section 206(d) (2) and (4) of this title [42 U.S. C § 2486d(d)(2) and (4)] and to such other terms, conditions, and requirements as the Administrator considers necessary or appropriate.

SEC. 2486f. SPACE GRANT COLLEGE AND SPACE GRANT REGIONAL CONSORTIUM

(a) Designation qualifications

(1) The Administrator may designate–

(A) any institution of higher education as a space grant college; and

(B) any association or other alliance of two or more persons, other than individuals, as a space grant regional consortium.

(2) No institution of higher education may be designated as a space grant college, unless the Administrator finds that such institution

(A) is maintaining a balanced program of research, education, training, and advisory services in fields related to space;

(B) will act in accordance with such guidelines as are prescribed under subsection (b)(2) of this section; and

(C) meets such other qualifications as the Administrator considers necessary or appropriate.

(3) No association or other alliance of two or more persons may be designated as a space grant regional consortium, unless the Administrator finds that such association or alliance–

(A) is established for the purpose of sharing expertise, research, educational facilities or training facilities, and other capabilities in order to facilitate research, education, training, and advisory services, in any field related to space;

(B) will encourage and follow a regional approach to solving problems or meeting needs relating to space, in cooperation with appropriate space grant colleges, space grant programs, and other persons in the region;

(C) will act in accordance with such guidelines as are prescribed under subsection (b)(2) of this section; and

(D) meets such other qualifications as the Administrator considers necessary or appropriate.

(b) Other necessary qualifications and guidelines on activities and responsibilities; regulations

The Administrator shall by regulation prescribe–

(1) the qualifications required to be met under subsection (a)(2)(C) and (3)(D) of this section; and

(2) guidelines relating to the activities and responsibilities of space grant colleges and space grant regional consortia.

(c) Suspension or termination of designation; hearing

The Administrator may, for cause and after an opportunity for hearing, suspend or terminate any designation under subsection (a) of this section.

SEC. 2486g. SPACE GRANT FELLOWSHIP PROGRAM

(a) Award of fellowships; guidelines; wide geographic and institutional diversity

The Administrator shall support a space grant fellowship program to provide educational and training assistance to qualified individuals at the graduate level of education in fields related to space. Such fellowships shall be awarded pursuant to guidelines established by the Administrator. Space grant fellowships shall be awarded to individuals at space grant colleges, space grant regional consortia, other colleges and institutions of higher education, professional associations, and institutes in such a manner as to assure wide geographic and institutional diversity in the pursuit of research under the fellowship program.

(b) Limitation on amount to provide grants

The total amount which may be provided for grants under the space grant fellowship program during any fiscal year shall not exceed an amount equal to 50 percent of the total funds appropriated for such year pursuant to this title.

(c) Authority to sponsor other research fellowship programs unaffected

Nothing in this section shall be construed to prohibit the Administrator from sponsoring any research fellowship program, including any special emphasis program, which is established under an authority other than this title.

SEC. 2486h. SPACE GRANT REVIEW PANEL

(a) Establishment

The Administrator shall establish an independent committee known as the space grant review panel, which shall not be subject to the provisions of the Federal Advisory Committee Act (5 U.S. C App. 1 *et seq.; Public Law 92-463).*

(b) Duties

The panel shall take such steps as may be necessary to review, and shall advise the Administrator with respect to–

(1) applications or proposals for, and performance under, grants and contracts awarded pursuant to sections 206 and 207 of this title (42 U.S. C §§ 2486d *and 2486e);*

(2) the space grant fellowship program;

(3) the designation and operation of space grant colleges and space grant regional consortia, and the operation of space grant and fellowship programs;

(4) the formulation and application of the planning guidelines and priorities pursuant to section 205(a) and (b)(1) of this title *[42 US. C § 2486c(a) and (b)(1)];* and

(5) such other matters as the Administrator refers to the panel for review and advice.

(c) Personnel and administrative services

The Administrator shall make available to the panel any information, personnel and administrative services and assistance which is reasonable to carry out the duties of the panel.

(d) Appointment of voting members; Chairman and Vice Chairman; reimbursement of non-Federal employee members; meetings; powers

(1) The Administrator shall appoint the voting members of the panel. A majority of the voting members shall be individuals who, by reason of knowledge, experience, or training, are especially qualified in one or more of the disciplines and fields related to space. The other voting members shall be individuals who, by reason of knowledge, experience or training, are especially qualified in, or representative of, education, extension services, State government, industry, economics, planning, or any other activity related to efforts to enhance the understanding, assessment, development, or utilization of space resources. The Administrator shall consider the potential conflict of interest of any individual in making appointments to the panel.

(2) The Administrator shall select one voting member to serve as the Chairman and another voting member to serve as the Vice Chairman. The Vice Chairman shall act as Chairman in the absence or incapacity of the Chairman.

(3) Voting members of the panel who are not Federal employees shall be reimbursed for actual and reasonable expenses incurred in the performance of such duties.

(4) The panel shall meet on a biannual basis and, at any other time, at the call of the Chairman or upon the request of a majority of the voting members or of the Administrator.

(5) The panel may exercise such powers as are reasonably necessary in order to carry out the duties enumerated in subsection (b) of this section.

SEC. 2486i. AVAILABILITY OF OTHER FEDERAL PERSONNEL AND DATA; COOPERATION WITH ADMINISTRATION

Each department, agency or other instrumentality of the Federal Government which is engaged in or concerned with, or which has authority over, matters relating to space

(1) may, upon a written request from the Administrator, make available, on a reimbursable basis or otherwise, any personnel (with their consent and without prejudice to their position and rating), service, or facility which the Administrator considers necessary to carry out any provision of this title;

(2) may, upon a written request from the Administrator, furnish any available data or other information which the Administrator considers necessary to carry out any provision of this title; and

(3) may cooperate with the Administration.

SEC. 2486j. REPORTS TO CONGRESS AND PRESIDENT; COMMENTS AND RECOMMENDATIONS

[Repealed by Pub. L. No. 105-362, Title XL § 1101 (a), 112 Stat. 3280, 3292 (Nov. 10, 1998).]

SEC. 2486k. DESIGNATION OR AWARD TO BE ON COMPETITIVE BASIS

The Administrator shall not under this title designate any space grant college or space grant regional consortium or award any fellowship, grant, or contract unless such designation or award is made in accordance with the competitive, merit-based review process employed by the Administration on the date of enactment of this Act.

SEC. 2486l. AUTHORIZATION OF APPROPRIATIONS

(a) There are authorized to be appropriated for the purposes of carrying out the provisions of this title sums not to exceed–

(1) $10,000,000 for each of fiscal years 1988 and 1989; and

(2) $15,000,000 for each of fiscal years 1990 and 1991.

(b) Such sums as may be appropriated under this section shall remain available until expended.

15 U.S.C. § 3710
Utilization of Federal Technology-Cooperative Research and Development Agreements (CRDAs)

(a) Policy

(1) It is the continuing responsibility of the Federal Government to ensure the full use of the results of the Nation's Federal investment in research and development. To this end the Federal Government shall strive where appropriate to transfer federally owned or originated technology to State and local governments and to the private sector.

(2) Technology transfer, consistent with mission responsibilities, is a responsibility of each laboratory science and engineering professional.

(3) Each laboratory director shall ensure that efforts to transfer technology are considered positively in laboratory job descriptions, employee promotion policies, and evaluation of the job performance of scientists and engineers in the laboratory.

(b) Establishment of Research and Technology Applications Offices

Each Federal laboratory shall establish an Office of Research and Technology Applications. Laboratories having existing organizational structures which perform the functions of this section may elect to combine the Office of Research and Technology Applications within the existing organization. The staffing and funding levels for these offices shall be determined between each Federal laboratory and the Federal agency operating or directing the laboratory, except that (1) each laboratory having 200 or more full-time equivalent scientific, engineering, and related technical positions shall provide one or more full-time equivalent positions as staff for its Office of Research and Technology Applications, and (2) each Federal agency which operates or directs one or more Federal laboratories shall make available sufficient funding, either as a separate line item or from the agency's research and development budget, to support the technology transfer function at the agency and at its laboratories, including support of the Offices of Research and Technology Applications.

Furthermore, individuals filling positions in an Office of Research and Technology Applications shall be included in the overall laboratory/agency management development program so as to ensure that highly competent technical managers are full participants in the technology transfer process.

The agency head shall submit to Congress at the time the President submits the budget to Congress an explanation of the agency's technology transfer program for the preceding year and the agency's plans for conducting its technology transfer function for the upcoming year, including plans for securing intellectual property rights in laboratory innovations with commercial promise and plans for managing such innovations so as to benefit the competitiveness of United States industry.

(c) Functions of Research and Technology Applications Offices

It shall be the function of each Office of Research and Technology Applications

(1) to prepare application assessments for selected research and development projects in which that laboratory is engaged and which in the opinion of the laboratory may have potential commercial applications;

(2) to provide and disseminate information on federally owned or originated products, processes, and services having potential application to State and local governments and to private industry;

(3) to cooperate with and assist the National Technical Information Service, the Federal Laboratory Consortium for Technology Transfer, and other organizations which link the research and development resources of that laboratory and the Federal Government as a whole to potential users in State and local government and private industry;

(4) to provide technical assistance to State and local government officials; and

(5) to participate, where feasible, in regional, State, and local programs designed to facilitate or stimulate the transfer of technology for the benefit of the region, State, or local jurisdiction in which the Federal laboratory is located. Agencies which have established organizational structures outside their Federal laboratories, which have as their principal purpose the transfer of federally owned or originated technology to State and local government and to the private sector may elect to perform the functions of this subsection in such organizational structures. No Office of Research and Technology Applications or other organizational structures performing the functions of this subsection shall substantially compete with similar services available in the private sector.

(d) Dissemination of technical information

The National Technical Information Service shall

(1) serve as a central clearinghouse for the collection, dissemination and transfer of information on federally owned or originated technologies having potential application to State and local governments and to private industry;

(2) utilize the expertise and services of the National Science Foundation and the Federal Laboratory Consortium for Technology Transfer, particularly in dealing with State and local governments;

(3) receive requests for technical assistance from State and local governments, respond to such requests with published information available to the Service, and refer such requests to the Federal Laboratory Consortium for Technology Transfer to the extent that such requests require a response involving more than the published information available to the Service;

(4) provide funding, at the discretion of the Secretary, for Federal laboratories to provide the assistance specified in subsection (c) (3) of this section;

(5) use appropriate technology transfer mechanisms such as personnel exchanges and computer-based systems; and

(6) maintain a permanent archival repository and clearinghouse for the collection and dissemination of nonclassified scientific, technical, and engineering information.

(e) Establishment of Federal Laboratory Consortium for Technology Transfer

(1) There is hereby established the Federal Laboratory Consortium for Technology Transfer (hereinafter referred to as the "Consortium') which, in cooperation with Federal laboratories and the private sector, shall

(A) develop and (with the consent of the Federal laboratory concerned) administer techniques, training courses, and materials concerning technology transfer to increase the awareness of Federal laboratory employees regarding the commercial potential of laboratory technology and innovations;

(B) furnish advice and assistance requested by Federal agencies and laboratories for use in their technology transfer programs (including the planning of seminars for small business and other industry);

(C) provide a clearinghouse for requests, received at the laboratory level, for technical assistance from States and units of local governments, businesses, industrial development organizations, not-for-profit organizations including universities, Federal agencies and laboratories, and other persons, and (i) to the extent that such requests can be responded to with published information available to the National Technical Information Service, refer such requests to that Service, and (ii) otherwise refer these requests to the appropriate Federal laboratories and agencies;

(D) facilitate communication and coordination between Offices of Research and Technology Applications of Federal laboratories;

(E) utilize (with the consent of the agency involved) the expertise and services of the National Science Foundation, the Department of Commerce, the National Aeronautics and Space Administration, and other Federal agencies, as necessary;

(F) with the consent of any Federal laboratory, facilitate the use by such laboratory of appropriate technology transfer mechanisms such as personnel exchanges and computer-based systems;

(G) with the consent of any Federal laboratory, assist such laboratory to establish programs using technical volunteers to provide technical assistance to communities related to such laboratory;

(H) facilitate communication and cooperation between Offices of Research and Technology Applications of Federal laboratories and regional, State, and local technology transfer organizations;

(I) when requested, assist colleges or universities, businesses, nonprofit organizations, State or local governments, or regional organizations to establish programs to stimulate research and to encourage technology transfer in such areas as technology program development, curriculum design, long-term research planning, personnel needs projections, and productivity assessments;

(J) seek advice in each Federal laboratory consortium region from representatives of State and local governments, large and small business, universities, and other appropriate persons on the effectiveness of the program (and any such advice shall be provided at no expense to the Government); and

(K) work with the Director of the National Institute on Disability and Rehabilitation Research to compile a compendium of current and projected Federal Laboratory technologies and projects that have or will have an intended or recognized impact on the available range of assistive technology for individuals with disabilities (as defined in section 3002 of Title 29), including technologies and projects that incorporate the principles of universal design (as defined in section 3002 of Title 29), as appropriate.

(2) The membership of the Consortium shall consist of the Federal laboratories described in clause (1) of subsection (b) of this section and such other laboratories as may choose to join the Consortium. The representatives to the Consortium shall include a senior staff member of each Federal laboratory which is a member of the Consortium and a senior representative appointed from each Federal agency with one or more member laboratories.

(3) The representatives to the Consortium shall elect a Chairman of the Consortium.

(4) The Director of the National Institute of Standards and Technology shall provide the Consortium, on a reimbursable basis, with administrative services, such as office space, personnel, and support services of the Institute, as requested by the Consortium and approved by such Director.

(5) Each Federal laboratory or agency shall transfer technology directly to users or representatives of users, and shall not transfer technology directly to the Consortium. Each Federal laboratory shall conduct and transfer technology only in accordance with the practices and policies of the Federal agency which owns, leases, or otherwise uses such Federal laboratory.

(6) Not later than one year after October 20, 1986, and every year thereafter, the Chairman of the Consortium shall submit a report to the President, to the appropriate authorization and appropriation committees of both Houses of the Congress,

and to each agency with respect to which a transfer of funding is made (for the fiscal year or years involved) under paragraph (7), concerning the activities of the Consortium and the expenditures made by it under this subsection during the year for which the report is made. Such report shall include an annual independent audit of the financial statements of the Consortium, conducted in accordance with generally accepted accounting principles.

(7)(A) Subject to subparagraph (B), an amount equal to 0.008 percent of the budget of each Federal agency from any Federal source, including related overhead, that is to be utilized by or on behalf of the laboratories of such agency for a fiscal *year* referred to in subparagraph (B)(ii) shall be transferred by such agency to the National Institute of Standards and Technology at the beginning of the fiscal year involved. Amounts so transferred shall be provided by the Institute to the Consortium for the purpose of carrying out activities of the Consortium under this subsection.

(B) A transfer shall be made by any Federal agency under subparagraph (A), for any fiscal year, only if the amount so transferred by that agency (as determined under such subparagraph) would exceed $10,000.

(C) The heads of Federal agencies and their designees, and the directors of Federal laboratories, may provide such additional support for operations of the Consortium as they deem appropriate.

(f) Repealed. Pub. L. 104-66, Title III, § 3001(f), Dec. 21, 1995, 109 Stat. 734.

(g) Functions of Secretary

(1) The Secretary, through the Under Secretary, and in consultation with other Federal agencies, may

(A) make available to interested agencies the expertise of the Department of Commerce regarding the commercial potential of inventions and methods and options for commercialization which are available to the Federal laboratories, including research and development limited partnerships;

(B) develop and disseminate to appropriate agency and laboratory personnel model provisions for use on a voluntary basis in cooperative research and development arrangements; and

(C) furnish advice and assistance, upon request, to Federal agencies concerning their cooperative research and development programs and projects.

(2) Two years after October 20, 1986 and every two years thereafter, the Secretary shall submit a summary report to the President and the Congress on the use by the agencies and the Secretary of the authorities specified in this chapter. Other Federal agencies shall cooperate in the report's preparation.

(3) Not later than one year after October 20, 1986, the Secretary shall submit to the President and the Congress a report regarding–

(A) any copyright provisions or other types of barriers which tend to restrict or limit the transfer of federally funded computer software to the private sector and to State and local governments, and agencies of such State and local governments; and

(B) the feasibility and cost of compiling and maintaining a current and comprehensive inventory of all federally funded training software.
(h) Repealed. Pub. L. 100-519, Title II, § 212(a)(4), Oct. 24, 1988, 102 Star. 2595.
(i) Research equipment
The Director of a laboratory, or the head of any Federal agency or department, may loan, lease, or give research equipment that is excess to the needs of the laboratory, agency, or department to an educational institution or nonprofit organization for the conduct of technical and scientific education and research activities. Title of ownership shall transfer with a gift under the section.

15 U.S.C. § 5806
Anchor tenancy and termination liability

(a) Anchor tenancy contracts
Subject to appropriations, the Administrator or the Administrator of the National Oceanic and Atmospheric Administration may enter into multiyear anchor tenancy contracts for the purchase of a good or service if the appropriate Administrator determines that—

(1) the good or service meets the mission requirements of the National Aeronautics and Space Administration or the National Oceanic and Atmospheric Administration, as appropriate;

(2) the commercially procured good or service is cost effective;

(3) the good or service is procured through a competitive process;

(4) existing or potential customers for the good or service other than the United States Government have been specifically identified;

(5) the long-term viability of the venture is not dependent upon a continued Government market or other nonreimbursable Government support; and

(6) private capital is at risk in the venture.

(b) Termination liability

(1) Contracts entered into under subsection (a) of this section may provide for the payment of termination liability in the event that the Government terminates such contracts for its convenience.

(2) Contracts that provide for the payment of termination liability, as described in paragraph (1), shall include a fixed schedule of such termination liability payments. Liability under such contracts shall not exceed the total payments which the Government would have made after the date of termination to purchase the good or service if the contract were not terminated.

(3) Subject to appropriations, funds available for such termination liability payments may be used for purchase of the good or service upon successful delivery of the good or service pursuant to the contract. In such case, sufficient funds shall remain available to cover any remaining termination liability.

(c) Limitations

(1) Contracts entered into under this section shall not exceed 10 years in duration.

(2) Such contracts shall provide for delivery of the good or service on a firm, fixed price basis.

(3) To the extent practicable, reasonable performance specifications shall be used to define technical requirements in such contracts.

(4) In any such contract, the appropriate Administrator shall reserve the right to completely or partially terminate the contract without payment of such termination liability because of the contractor's actual or anticipated failure to perform its contractual obligations.

15 U.S.C. § 5807
Use of Government facilities

(a) Authority Federal agencies, including the National Aeronautics and Space Administration and the Department of Defense, may allow non-Federal entities to use their space-related facilities on a reimbursable basis if the Administrator, the Secretary of Defense, or the appropriate agency head determines that

(1) the facilities will be used to support commercial space activities;

(2) such use can be supported by existing or planned Federal resources;

(3) such use is compatible with Federal activities;

(4) equivalent commercial services are not available on reasonable terms; and

(5) such use is consistent with public safety, national security, and international treaty obligations. In carrying out paragraph (5), each agency head shall consult with appropriate Federal officials.

(b) Reimbursement payment

(1) The reimbursement referred to in subsection (a) of this section may be an amount equal to the direct costs (including salaries of United States civilian and contractor personnel) incurred by the United States as a result of the use of such facilities by the private sector. For the purposes of this paragraph, the term "direct costs" means the actual costs that can be unambiguously associated with such use, and would not be borne by the United States Government in the absence of such use.

(2) The amount of any payment received by the United States for use of facilities under this subsection shall be credited to the appropriation from which the cost of providing such facilities was paid.

18 U.S.C. § 7
Special maritime and territorial jurisdiction of the United States defined

The term "special maritime and territorial jurisdiction of the United States", as used in this title, includes:

(1) The high seas, any other waters within the admiralty and maritime jurisdiction of the United States and out of the jurisdiction of any particular State, and any vessel belonging in whole or in part to the United States or any citizen thereof, or to any corporation created by or under the laws of the United States, or of any State, Territory, District, or possession thereof, when such vessel is within the admiralty and maritime jurisdiction of the United States and out of the jurisdiction of any particular State.

(2) Any vessel registered, licensed, or enrolled under the laws of the United States, and being on a voyage upon the waters of any of the Great Lakes, or any of the waters connecting them, or upon the Saint Lawrence River where the same constitutes the International Boundary Line.

(3) Any lands reserved or acquired for the use of the United States, and under the exclusive or concurrent jurisdiction thereof, or any place purchased or otherwise acquired by the United States by consent of the legislature of the State in which the same shall be, for the erection of a fort, magazine, arsenal, dockyard, or other needful building.

(4) Any island, rock, or key containing deposits of guano, which may, at the discretion of the President, be considered as appertaining to the United States.

(5) *Any* aircraft belonging in whole or in part to the United States, or any citizen thereof, or to any corporation created by or under the laws of the United States, or any State, Territory, district, or possession thereof, while such aircraft is in flight over the high seas, or over *any* other waters within the admiralty and maritime jurisdiction of the United States and out of the jurisdiction of any particular State.

(6) Any vehicle used or designed for flight or navigation in space and on the registry of the United States pursuant to the Treaty on Principles Governing the Activities of States in the Exploration and Use of Outer Space, Including the Moon and Other Celestial Bodies and the Convention on Registration of Objects Launched into Outer Space, while that vehicle is in flight, which is from the moment when all external doors are closed on Earth following embarkation until the moment when one such door is opened on Earth for disembarkation or in the case of a forced landing, until the competent authorities take over the responsibility for the vehicle and for persons and property aboard.

(7) Any place outside the jurisdiction of any nation with respect to an offense by or against a national of the United States.

(8) To the extent permitted by international law, any foreign vessel during a voyage having a scheduled departure from or arrival in the United States with respect to an offense committed by or against a national of the United States.

18 U.S.C. § 1905
Disclosure of confidential information generally

Whoever, being an officer or employee of the United States or of any department or agency thereof, any person acting on behalf of the Office of Federal Housing Enterprise Oversight, or agent of the Department of justice as defined in the Antitrust Civil Process Act (15 U.S.C. 1311-1314), publishes, divulges, discloses, or makes known in any manner or to any extent not authorized by law any information coming to him in the course of his employment or official duties or by reason of any examination or investigation made by, or return, report or record made to or filed with, such department or agency or officer or employee thereof, which information concerns or relates to the trade secrets, processes, operations, style of work, or apparatus, or to the identity, confidential statistical data, amount or source of any income, profits, losses, or expenditures of any person, firm, partnership, corporation, or association; or permits any income return or copy thereof or any book containing any abstract or particulars thereof to be seen or examined by any person except as provided by law; shall be fined under this title, or imprisoned not more than one year, or both; and shall be removed from office or employment.

35 U.S.C. § 105
Inventions in outer space

(a) Any invention made, used or sold in outer space on a space object or component thereof under the jurisdiction or control of the United States shall be considered to be made, used or sold within the United States for the purposes of this title, except with respect to any space object or component thereof that is specifically identified and otherwise provided for by an international agreement to which the United States is a party, or with respect to any space object or component thereof that is carried on the registry of a foreign state in accordance with the Convention on Registration of Objects Launched into Outer Space.
(b) Any invention made, used or sold in outer space on a space object or component thereof that is carried on the registry of a foreign state in accordance with the Convention on Registration of Objects Launched into Outer Space, shall be considered to be made, used or sold within the United States for the purposes of this title if specifically so agreed in an international agreement between the United States and the state of registry.

42 U.S.C. § 2459d
Prohibition of grant or contract providing guaranteed customer base for new commercial space hardware or services

No amount appropriated to the National Aeronautics and Space Administration in this or any other Act with respect to any fiscal year may be used to fund grants, contracts or other agreements with an expected duration of more than one year, when a primary effect of the grant, contract, or agreement is to provide a guaranteed customer base for or establish an anchor tenancy in new commercial space hardware or services unless an appropriations Act specifies the new commercial space hardware or services to be developed or used, or the grant, contract, or agreement is otherwise identified in such Act.

42 U.S.C. § 2464
Recovery of fair value of placing Department of Defense payloads in orbit with Space Shuttle

Notwithstanding any other provision of law, or any interagency agreement, the Administrator of the National Aeronautics and Space Administration shall charge such prices as necessary to recover the fair value of placing Department of Defense payloads into orbit by means of the Space Shuttle.

42 U.S.C. § 2465a
Space Shuttle use policy

(a) Use policy

(1) It shall be the policy of the United States to use the Space Shuttle for purposes that (i) require the presence of man, (ii) require the unique capabilities of the Space Shuttle or (iii) when other compelling circumstances exist.

(2) The term "compelling circumstances" includes, but is not limited to, occasions when the Administrator determines, in consultation with the Secretary of Defense and the Secretary of State, that important national security or foreign policy interests would be served by a Shuttle launch.

(3) The policy stated in subsection (a) (1) of this section shall not preclude the use of available cargo space, on a Space Shuttle mission otherwise consistent with the policy described under subsection (a)(1) of this section, for the purpose of carrying secondary payloads (as defined by the Administrator) that do not require the presence of man if such payloads are consistent with the requirements of research, development, demonstration, scientific, commercial, and educational programs authorized by the Administrator.

(b) Implementation plan

The Administrator shall, within six months after November 16, 1990, submit a report to the Congress setting forth a plan for the implementation of the policy described in subsection (a)(1) of this section. Such plan shall include

(1) details of the implementation plan;

(2) a list of purposes that meet such policy;

(3) a proposed schedule for the implementation of such policy;

(4) an estimate of the costs to the United States of implementing such policy; and

(5) a process for informing the Congress in a timely and regular manner of how the plan is being implemented.

(c) Annual report

At least annually, the Administrator shall submit to the Congress a report certifying that the payloads scheduled to be launched on the space shuttle for the next four years are consistent with the policy set forth in subsection (a) (1) of this section. For each payload scheduled to be launched from the space shuttle which do not require the presence of man, the Administrator shall, in the certified report to Congress, state the specific circumstances which justified the use of the space shuttle. If, during the period between scheduled reports to the Congress, any additions are made to the list of certified payloads intended to be launched from the Shuttle, the Administrator shall inform the Congress of the additions and the reasons therefor within 45 days of the change.

(d) NASA payloads

The report described in subsection (c) of this section shall also include those National Aeronautics and Space Administration payloads designed solely to fly on the space shuttle which have begun the phase C/D of its development cycle.

42 U.S.C. § 2465c
Definitions

For the purposes of this title–

(1) the term "launch vehicle" means any vehicle constructed for the purpose of operating in, or placing a payload in, outer space; and

(2) the term "payload" means an object which a person undertakes to place in outer space by means of a launch vehicle, and includes sub-components of the launch vehicle specifically designed or adapted for that object.

42 U.S.C. § 2465f
Other activities of National Aeronautics and Space Administration

Commercial payloads may not be accepted for launch as primary payloads on the space shuttle unless the Administrator of the National Aeronautics and Space Administration determines that–

(1) the payload requires the unique capabilities of the space shuttle; or

(2) launching of the payload on the space shuttle is important for either national security or foreign policy purposes.

42 U.S.C. § 2466
Shuttle pricing policy; Congressional findings and declaration of purpose

The Congress finds and declares that–

(1) the Space Transportation System is a vital element of the United States space program, contributing to the United States leadership in space research, technology, and development;

(2) the Space Transportation System is the primary space launch system for both United States national security and civil government missions;

(3) the Space Transportation System contributes to the expansion of United States private sector investment and involvement in space and therefore should serve commercial users;

(4) the availability of the Space Transportation System to foreign users for peaceful purposes is an important means of promoting international cooperative activities in the national interest and in maintaining access to space for activities which enhance the security and welfare of mankind;

(5) the United States is committed to maintaining world leadership in space transportation;

(6) making the Space Transportation System fully operational and cost effective in providing routine access to space will maximize the national economic benefits of the system; and

(7) national goals and the objectives for the Space Transportation System can be furthered by a stable and fair pricing policy for the Space Transportation System.

42 U.S.C. § 14713
Acquisition of space science data

(a) Acquisition from commercial providers
The Administrator shall, to the extent possible and while satisfying the scientific or educational requirements of the National Aeronautics and Space Administration, and where appropriate, of other Federal agencies and scientific researchers, acquire, where cost effective, space science data from a commercial provider.
(b) Treatment of space science data as commercial item under acquisition laws
Acquisitions of space science data by the Administrator shall be carried out in accordance with applicable acquisition laws and regulations (including chapters 137 and 140 of title 10, United States Code). For purposes of such law and regulations, space science data shall be considered to be a commercial item. Nothing in this subsection

shall be construed to preclude the United States from acquiring, through contracts with commercial providers, sufficient rights in data to meet the needs of the scientific and educational community or the needs of other government activities.

(c) Definition

For purposes of this section, the term "space science data" includes scientific data concerning—

(1) the elemental and mineralogical resources of the moon, asteroids, planets and their moons, and comets;

(2) microgravity acceleration; and

(3) solar storm monitoring.

(d) Safety standards

Nothing in this section shall be construed to prohibit the Federal Government from requiring compliance with applicable safety standards.

(e) Limitation

This section does not authorize the National Aeronautics and Space Administration to provide financial assistance for the development of commercial systems for the collection of space science data.

42 U.S.C. § 14715
Sources of earth science data

(a) Acquisition

The Administrator shall, to the extent possible and while satisfying the scientific or educational requirements of the National Aeronautics and Space Administration, and where appropriate, of other Federal agencies and scientific researchers, acquire, where cost-effective, space-based and airborne Earth remote sensing data, services, distribution, and applications from a commercial provider.

(b) Treatment as commercial item under acquisition laws

Acquisitions by the Administrator of the data, services, distribution, and applications referred to in subsection (a) shall be carried out in accordance with applicable acquisition laws and regulations (including chapters 137 and 140 of tide 10, United States Code). For purposes of such law and regulations, such data, services, distribution, and applications shall be considered to be a commercial item. Nothing in this subsection shall be construed to preclude the United States from acquiring, through contracts with commercial providers, sufficient rights in data to meet the needs of the scientific and educational community or the needs of other government activities.

(c) Study

(1) The Administrator shall conduct a study to determine the extent to which the baseline scientific requirements of Earth Science can be met by commercial provid-

ers, and how the National Aeronautics and Space Administration will meet such requirements which cannot be met by commercial providers.

(2) The study conducted under this subsection shall—

(A) make recommendations to promote the availability of information from the National Aeronautics and Space Administration to commercial providers to enable commercial providers to better meet the baseline scientific requirements of Earth Science;

(B) make recommendations to promote the dissemination to commercial providers of information on advanced technology research and development performed by or for the National Aeronautics and Space Administration; and

(C) identify policy, regulatory, and legislative barriers to the implementation of the recommendations made under this subsection.

(3) The results of the study conducted under this subsection shall be transmitted to the Congress within 6 months after the date of the enactment of this Act.

(d) Safety standards

Nothing in this section shall be construed to prohibit the Federal Government from requiring compliance with applicable safety standards.

(e) Administration and execution

This section shall be carried out as part of the Commercial Remote Sensing Program at the Stennis Space Center.

(f) [Omitted] (Pub. L. 105-303, Title I § 107, Oct. 28, 1998, 112 Stat. 2853.)

Commission on the Future of the United States Aerospace Industry
Pub. L. 106–398, § 1 [[div. A], title X, § 1092], Oct. 30, 2000, 114 Stat. 1654, 1654A–300, as amended by Pub. L. 107–107, div. A, title X, § 1062, Dec. 28, 2001, 115 Stat. 1232
Codified at 42 U.S.C. § 2451 note

SEC. 2451

(a) Establishment.—There is established a commission to be known as the 'Commission on the Future of the United States Aerospace Industry' (in this section referred to as the 'Commission').

(b) Membership.—

(1) The Commission shall be composed of 12 members appointed, not later than March 1, 2001, as follows:

(A) Up to six members shall be appointed by the President.

(B) Two members shall be appointed by the Speaker of the House of Representatives.

(C) Two members shall be appointed by the majority leader of the Senate.

(D) One member shall be appointed by the minority leader of the Senate.

(E) One member shall be appointed by the minority leader of the House of Representatives.

(2) The members of the Commission shall be appointed from among persons with extensive experience and national reputations in aerospace manufacturing, economics, finance, national security, international trade, or foreign policy and persons who are representative of labor organizations associated with the aerospace industry.

(3) Members shall be appointed for the life of the Commission. A vacancy in the Commission shall not affect its powers, but shall be filled in the same manner as the original appointment.

(4) The President shall designate one member of the Commission to serve as the chairman of the Commission.

(5) The Commission shall meet at the call of the chairman. A majority of the members shall constitute a quorum, but a lesser number may hold hearings.

(c) Duties.—

(1) The Commission shall—

(A) study the issues associated with the future of the United States aerospace industry in the global economy, particularly in relationship to United States national security; and

(B) assess the future importance of the domestic aerospace industry for the economic and national security of the United States.

(2) In order to fulfill its responsibilities, the Commission shall study the following:

(A) The budget process of the United States Government, particularly with a view to assessing the adequacy of projected budgets of the Federal departments and agencies for aerospace research and development and procurement.

(B) The acquisition process of the Government, particularly with a view to assessing—

(i) the adequacy of the current acquisition process of Federal departments and agencies; and

(ii) the procedures for developing and fielding aerospace systems incorporating new technologies in a timely fashion.

(C) The policies, procedures, and methods for the financing and payment of Government contracts.

(D) Statutes and regulations governing international trade and the export of technology, particularly with a view to assessing—

(i) the extent to which the current system for controlling the export of aerospace goods, services, and technologies reflects an adequate balance between the need to protect national security and the need to ensure unhindered access to the global marketplace; and

(ii) the adequacy of United States and multilateral trade laws and policies for maintaining the international competitiveness of the United States aerospace industry.

(E) Policies governing taxation, particularly with a view to assessing the impact of current tax laws and practices on the international competitiveness of the aerospace industry.

(F) Programs for the maintenance of the national space launch infrastructure, particularly with a view to assessing the adequacy of current and projected programs for maintaining the national space launch infrastructure.

(G) Programs for the support of science and engineering education, including current programs for supporting aerospace science and engineering efforts at institutions of higher learning, with a view to determining the adequacy of those programs.

(d) Report.—

(1) Not later than one year after the date of the first official meeting of the Commission, the Commission shall submit a report on its activities to the President and Congress.

(2) The report shall include the following:

(A) The Commission's findings and conclusions.

(B) The Commission's recommendations for actions by Federal departments and agencies to support the maintenance of a robust aerospace industry in the United States in the 21st century and any recommendations for statutory and regulatory changes to support the implementation of the Commission's findings.

(C) A discussion of the appropriate means for implementing the Commission's recommendations.

(e) Administrative Requirements and Authorities.–

(1) The Director of the Office of Management and Budget shall ensure that the Commission is provided such administrative services, facilities, staff, and other support services as may be necessary. Any expenses of the Commission shall be paid from funds available to the Director.

(2) The Commission may hold hearings, sit and act at times and places, take testimony, and receive evidence that the Commission considers advisable to carry out the purposes of this section.

(3) The Commission may request directly from any department or agency of the United States any information that the Commission considers necessary to carry out the provisions of this section. To the extent consistent with applicable requirements of law and regulations, the head of such department or agency shall furnish such information to the Commission.

(4) The Commission may use the United States mails in the same manner and under the same conditions as other departments and agencies of the United States.

(f) Commission Personnel Matters.—

(1) Members of the Commission shall serve without additional compensation for their service on the Commission, except that members appointed from among private citizens may be allowed travel expenses, including per diem in lieu of subsistence, as authorized by law for persons serving intermittently in Government service

under subchapter I of chapter 57 of title 5, United States Code, while away from their homes and places of business in the performance of services for the Commission.

(2) The chairman of the Commission may appoint staff of the Commission, request the detail of Federal employees, and accept temporary and intermittent services in accordance with section 3161 of title 5, United States Code (as added by section 1101 of this Act).

(g) Termination.—The Commission shall terminate 60 days after the date of the submission of its report under subsection (d)."

International Space Station Contingency Plan
P.L. 106-391 Title II, 114 Stat. 1586 (Oct. 30, 2000)
Codified at 42 U.S.C. § 2451 note

SEC. 2451. International Space Station Contingency Plan.

(a) Bimonthly Reporting on Russian Status.—Not later than the first day of the first month beginning more than 60 days after the date of the enactment of this Act [Oct. 30, 2000], and not later than the first day of every second month thereafter until October 1, 2006, the Administrator [of the National Aeronautics and Space Administration] shall report to Congress whether or not the Russians have performed work expected of them and necessary to complete the International Space Station. Each such report shall also include a statement of the Administrator's judgment concerning Russia's ability to perform work anticipated and required to complete the International Space Station before the next report under this subsection.

(b) Decision on Russian Critical Path Items.—The President shall notify Congress within 90 days after the date of the enactment of this Act [Oct. 30, 2000] of the decision on whether or not to proceed with permanent replacement of any Russian elements in the critical path [as defined in section 3 of Pub. L. 106–391, 42 U.S.C. 2452 note] of the International Space Station or any Russian launch services. Such notification shall include the reasons and justifications for the decision and the costs associated with the decision. Such decision shall include a judgment of when all elements identified in Revision E assembly sequence as of June 1999 will be in orbit and operational. If the President decides to proceed with a permanent replacement for any Russian element in the critical path or any Russian launch services, the President shall notify Congress of the reasons and the justification for the decision to proceed with the permanent replacement and the costs associated with the decision.

(c) Assurances.—The United States shall seek assurances from the Russian Government that it places a higher priority on fulfilling its commitments to the International Space Station than it places on extending the life of the Mir Space Station, including assurances that Russia will not utilize assets allocated by Russia to the International Space Station for other purposes, including extending the life of Mir.

(d) Equitable Utilization.—In the event that any International Partner in the Inter-

national Space Station Program willfully violates any of its commitments or agreements for the provision of agreed upon Space Station-related hardware or related goods or services, the Administrator should, in a manner consistent with relevant international agreements, seek a commensurate reduction in the utilization rights of that Partner until such time as the violated commitments or agreements have been fulfilled.
(e) Operation Costs.—The Administrator shall, in a manner consistent with relevant international agreements, seek to reduce the National Aeronautics and Space Administration's share of International Space Station common operating costs, based upon any additional capabilities provided to the International Space Station through the National Aeronautics and Space Administration's Russian Program Assurance activities.

COST LIMITATION FOR THE INTERNATIONAL SPACE STATION

(a) Limitation of Costs.—

(1) In general.—Except as provided in subsections (c) and (d), the total amount obligated by the National Aeronautics and Space Administration for—

(A) costs of the International Space Station may not exceed $25,000,000,000; and

(B) space shuttle launch costs in connection with the assembly of the International Space Station may not exceed $17,700,000,000.

(2) Calculation of launch costs.—For purposes of paragraph (1)(B)—

(A) not more than $380,000,000 in costs for any single space shuttle launch shall be taken into account; and

(B) if the space shuttle launch costs taken into account for any single space shuttle launch are less than $380,000,000, then the Administrator [of the National Aeronautics and Space Administration] shall arrange for a verification, by the General Accounting Office, of the accounting used to determine those costs and shall submit that verification to the Congress within 60 days after the date on which the next budget request is transmitted to the Congress.

(b) Costs to Which Limitation Applies.—

(1) Development costs.—The limitation imposed by subsection (a)(1)(A) does not apply to funding for operations, research, or crew return activities subsequent to substantial completion of the International Space Station.

(2) Launch costs.—The limitation imposed by subsection (a)(1)(B) does not apply—

(A) to space shuttle launch costs in connection with operations, research, or crew return activities subsequent to substantial completion of the International Space Station;

(B) to space shuttle launch costs in connection with a launch for a mission on which at least 75 percent of the shuttle payload by mass is devoted to research; nor

(C) to any additional costs incurred in ensuring or enhancing the safety and reliability of the space shuttle.

(3) Substantial completion.—For purposes of this subsection, the International Space Station is considered to be substantially completed when the development costs comprise 5 percent or less of the total International Space Station costs for the fiscal year.

(c) Notice of Changes to Space Station Costs.—The Administrator shall provide with each annual budget request a written notice and analysis of any changes under subsection (d) to the amounts set forth in subsection (a) to the Senate Committees on Appropriations and on Commerce, Science, and Transportation and to the House of Representatives Committees on Appropriations and on Science. In addition, such notice may be provided at other times, as deemed necessary by the Administrator. The written notice shall include—

(1) an explanation of the basis for the change, including the costs associated with the change and the expected benefit to the program to be derived from the change;

(2) an analysis of the impact on the assembly schedule and annual funding estimates of not receiving the requested increases; and

(3) an explanation of the reasons that such a change was not anticipated in previous program budgets.

(d) Funding for Contingencies.—

(1) Notice required.—If funding in excess of the limitation provided for in subsection (a) is required to address the contingencies described in paragraph (2), then the Administrator shall provide the written notice required by subsection (c). In the case of funding described in paragraph (3)(A), such notice shall be required prior to obligating any of the funding. In the case of funding described in paragraph (3)(B), such notice shall be required within 15 days after making a decision to implement a change that increases the space shuttle launch costs in connection with the assembly of the International Space Station.

(2) Contingencies.—The contingencies referred to in paragraph (1) are the following:

(A) The lack of performance or the termination of participation of any of the International countries party to the Intergovernmental Agreement.

(B) The loss or failure of a United States-provided element during launch or on-orbit.

(C) On-orbit assembly problems.

(D) New technologies or training to improve safety on the International Space Station.

(E) The need to launch a space shuttle to ensure the safety of the crew or to maintain the integrity of the station.

(3) Amounts.—The total amount obligated by the National Aeronautics and

Space Administration to address the contingencies described in paragraph (2) is limited to—

(A) $5,000,000,000 for the International Space Station; and

(B) $3,540,000,000 for the space shuttle launch costs in connection with the assembly of the International Space Station.

(e) Reporting and Review.—

(1) Identification of costs.—

(A) Space shuttle.—As part of the overall space shuttle program budget request for each fiscal year, the Administrator shall identify separately—

(i) the amounts of the requested funding that are to be used for completion of the assembly of the International Space Station; and

(ii) any shuttle research mission described in subsection (b)(2).

(B) International space station.—As part of the overall International Space Station budget request for each fiscal year, the Administrator shall identify the amount to be used for development of the International Space Station.

(2) Accounting for cost limitations.—As part of the annual budget request to the Congress, the Administrator shall account for the cost limitations imposed by subsection (a).

(3) Verification of accounting.—The Administrator shall arrange for a verification, by the General Accounting Office, of the accounting submitted to the Congress within 60 days after the date on which the budget request is transmitted to the Congress.

(4) Inspector general.—Within 60 days after the Administrator provides a notice and analysis to the Congress under subsection (c), the Inspector General of the National Aeronautics and Space Administration shall review the notice and analysis and report the results of the review to the committees to which the notice and analysis were provided.

RESEARCH ON INTERNATIONAL SPACE STATION

(a) Study.—The Administrator [of the National Aeronautics and Space Administration] shall enter into a contract with the National Research Council and the National Academy of Public Administration to jointly conduct a study of the status of life and microgravity research as it relates to the International Space Station. The study shall include—

(1) an assessment of the United States scientific community's readiness to use the International Space Station for life and microgravity research;

(2) an assessment of the current and projected factors limiting the United States scientific community's ability to maximize the research potential of the International Space Station, including, but not limited to, the past and present availability of resources in the life and microgravity research accounts within the Office of Human Spaceflight and the Office of Life and Microgravity Sciences and Applications and the past, present, and projected access to space of the scientific community; and

(3) recommendations for improving the United States scientific community's ability to maximize the research potential of the International Space Station, including an assessment of the relative costs and benefits of—

(A) dedicating an annual mission of the Space Shuttle to life and microgravity research during assembly of the International Space Station; and

(B) maintaining the schedule for assembly in place at the time of the enactment [Oct. 30, 2000].

(b) Report.—Not later than 1 year after the date of the enactment of this Act [Oct. 30, 2000], the Administrator shall transmit to the Committee on Science of the House of Representatives and the Committee on Commerce, Science, and Transportation of the Senate a report on the results of the study conducted under this section.

SPACE STATION RESEARCH UTILIZATION AND COMMERCIALIZATION MANAGEMENT

(a) Research Utilization and Commercialization Management Activities.—The Administrator of the National Aeronautics and Space Administration shall enter into an agreement with a non-government organization to conduct research utilization and commercialization management activities of the International Space Station subsequent to substantial completion as defined in section 202 (b)(3). The agreement may not take effect less than 120 days after the implementation plan for the agreement is submitted to the Congress under subsection (b).

(b) Implementation Plan.—Not later than September 30, 2001, the Administrator shall submit to the Committee on Commerce, Science, and Transportation of the Senate and the Committee on Science of the House of Representatives an implementation plan to incorporate the use of a non-government organization for the International Space Station. The implementation plan shall include—

(1) a description of the respective roles and responsibilities of the Administration and the non-government organization;

(2) a proposed structure for the non-government organization;

(3) a statement of the resources required;

(4) a schedule for the transition of responsibilities; and

(5) a statement of the duration of the agreement.

Aero-Space Transportation Technology Integration
Pub. L. 106–391, title III, § 308, Oct. 30, 2000, 114 Stat. 1592
Codified at 42 U.S.C. § 2451 note

SEC. 2451

(a) Integration Plan.—The Administrator [of the National Aeronautics and Space Administration] shall develop a plan for the integration of research, development, and experimental demonstration activities in the aeronautics transportation technology and space transportation technology areas where appropriate. The plan shall ensure that integration is accomplished without losing unique capabilities which support the National Aeronautics and Space Administration's defined missions. The plan shall also include appropriate strategies for using aeronautics centers in integration efforts.

(b) Reports to Congress.—Not later than 90 days after the date of the enactment of this Act [Oct. 30, 2000], the Administrator shall transmit to the Congress a report containing the plan developed under subsection (a). The Administrator shall transmit to the Congress annually thereafter for 5 years a report on progress in achieving such plan, to be transmitted with the annual budget request.

Innovative Technologies for Human Space Flight
Pub. L. 106–391, title III, § 313, Oct. 30, 2000, 114 Stat. 1594
Codified at 42 U.S.C. § 2451 note

SEC. 2451:

(a) Establishment of Program.—In order to promote a 'faster, cheaper, better' approach to the human exploration and development of space, the Administrator [of the National Aeronautics and Space Administration] shall establish a Human Space Flight Innovative Technologies program of groundbased and space-based research and development in innovative technologies. The program shall be part of the Technology and Commercialization program.

(b) Awards.—At least 75 percent of the amount appropriated for Technology and Commercialization under section 101 (b)(4) [114 Stat. 1581] for any fiscal year shall be awarded through broadly distributed announcements of opportunity that solicit proposals from educational institutions, industry, nonprofit institutions, National Aeronautics and Space Administration Centers, the Jet Propulsion Laboratory, other Federal agencies, and other interested organizations, and that allow partnerships among any combination of those entities, with evaluation, prioritization, and recommendations made by external peer review panels.

(c) Plan.—The Administrator shall provide to the Committee on Science of the House of Representatives and to the Committee on Commerce, Science, and Transportation of the Senate, not later than December 1, 2000, a plan to implement the program established under subsection (a).

Life in the Universe
Pub. L. 106–391, title III, § 314, Oct. 30, 2000, 114 Stat. 1595
Codified at 42 U.S.C. § 2451 note

SEC. 2451

(a) Review.—The Administrator [of the National Aeronautics and Space Administration] shall enter into appropriate arrangements with the National Academy of Sciences for the conduct of a review of—

(1) international efforts to determine the extent of life in the universe; and

(2) enhancements that can be made to the National Aeronautics and Space Administration's efforts to determine the extent of life in the universe.

(b) Elements.—The review required by subsection (a) shall include

(1) an assessment of the direction of the National Aeronautics and Space Administration's astrobiology initiatives within the Origins program;

(2) an assessment of the direction of other initiatives carried out by entities other than the National Aeronautics and Space Administration to determine the extent of life in the universe, including other Federal agencies, foreign space agencies, and private groups such as the Search for Extraterrestrial Intelligence Institute;

(3) recommendations about scientific and technological enhancements that could be made to the National Aeronautics and Space Administration's astrobiology initiatives to effectively utilize the initiatives of the scientific and technical communities; and

(4) recommendations for possible coordination or integration of National Aeronautics and Space Administration initiatives with initiatives of other entities described in paragraph (2).

(c) Report to Congress.—Not later than 20 months after the date of the enactment of this Act [Oct. 30, 2000], the Administrator shall transmit to the Congress a report on the results of the review carried out under this section.

Carbon Cycle Remote Sensing Applications Research
Pub. L. 106–391, title III, § 315, Oct. 30, 2000, 114 Stat. 1595
Codified at 42 U.S.C. § 2451 note

SEC. 2451

(a) Carbon Cycle Remote Sensing Applications Research Program

(1) In general.—The Administrator [of the National Aeronautics and Space Administration] shall develop a carbon cycle remote sensing applications research program—

(A) to provide a comprehensive view of vegetation conditions;

(B) to assess and model agricultural carbon sequestration; and

(C) to encourage the development of commercial products, as appropriate.

(2) Use of centers.—The Administrator of the National Aeronautics and Space Administration shall use regional earth science application centers to conduct applications research under this section.

(3) Researched areas.—The areas that shall be the subjects of research conducted under this section include—

(A) the mapping of carbon-sequestering land use and land cover;

(B) the monitoring of changes in land cover and management;

(C) new approaches for the remote sensing of soil carbon; and

(D) region-scale carbon sequestration estimation.

(b) Authorization of Appropriations.—There is authorized to be appropriated to carry out this section $5,000,000 of funds authorized by section 102 [114 Stat. 1581] for fiscal years 2001 through 2002.

100th Anniversary of Flight Educational Initiative
Pub. L. 106–391, title III, § 317, Oct. 30, 2000, 114 Stat. 1596
Codified at 42 U.S.C. § 2451 note

SEC. 2451

(a) Educational Initiative.—In recognition of the 100th anniversary of the first powered flight, the Administrator [of the National Aeronautics and Space Administration], in coordination with the Secretary of Education, shall develop and provide for the distribution, for use in the 2001–2002 academic year and thereafter, of age-appropriate educational materials, for use at the kindergarten, elementary, and secondary levels, on the history of flight, the contribution of flight to global development in the 20th century, the practical benefits of aeronautics and space flight to society, the scientific and mathematical principles used in flight, and any other related topics the Administrator considers appropriate. The Administrator shall integrate into the educational materials plans for the development and flight of the Mars plane.

(b) Report to Congress.—Not later than December 1, 2000, the Administrator shall transmit a report to the Congress on activities undertaken pursuant to this section.

National Aeronautics and Space Administration Authorization Act of 2000
Pub. L. 106–391, § 3, Oct. 30, 2000, 114 Stat. 1579
Codified at 42 U.S.C. § 2451 note

SEC. 2451. For purposes of this Act—

(1) the term 'Administrator' means the Administrator of the National Aeronautics and Space Administration;

(2) the term 'commercial provider' means any person providing space transporta-

tion services or other space-related activities, the primary control of which is held by persons other than a Federal, State, local, or foreign government;

(3) the term 'critical path' means the sequence of events of a schedule of events under which a delay in any event causes a delay in the overall schedule;

(4) the term 'grant agreement' has the meaning given that term in section 6302 (2) of title 31, United States Code;

(5) the term 'institution of higher education' has the meaning given such term in section 101 of the Higher Education Act of 1965 (20 U.S.C. 1001);

(6) the term 'State' means each of the several States of the United States, the District of Columbia, the Commonwealth of Puerto Rico, the Virgin Islands, Guam, American Samoa, the Commonwealth of the Northern Mariana Islands, and any other commonwealth, territory, or possession of the United States; and

(7) the term 'United States commercial provider' means a commercial provider, organized under the laws of the United States or of a State, which is—

(A) more than 50 percent owned by United States nationals; or

(B) a subsidiary of a foreign company and the Secretary of Commerce finds that—

(i) such subsidiary has in the past evidenced a substantial commitment to the United States market through—

(I) investments in the United States in long-term research, development, and manufacturing (including the manufacture of major components and subassemblies); and

(II) significant contributions to employment in the United States; and

(ii) the country or countries in which such foreign company is incorporated or organized, and, if appropriate, in which it principally conducts its business, affords reciprocal treatment to companies described in subparagraph (A) comparable to that afforded to such foreign company's subsidiary in the United States, as evidenced by—

(I) providing comparable opportunities for companies described in subparagraph (A) to participate in Government sponsored research and development similar to that authorized under this Act;

(II) providing no barriers to companies described in subparagraph (A) with respect to local investment opportunities that are not provided to foreign companies in the United States; and

(III) providing adequate and effective protection for the intellectual property rights of companies described in subparagraph (A).

Working Capital Fund
Pub. L. No. 108-7, Div K, Title III, 117 Stat. 520, on Feb. 20, 2003
Uncodified

There is hereby established in the United States Treasury a National Aeronautics and Space Administration working capital fund. Amounts in the fund are available for financing activities, services, equipment, information, and facilities as authorized by law to be provided within the Administration; to other agencies or instrumentalities of the United States; to any State, Territory, or possession or political subdivision thereof; to other public or private agencies; or to any person, firm, association, corporation, or educational institution on a reimbursable basis. The fund shall also be available for the purpose of funding capital repairs, renovations, rehabilitation, sustainment, demolition, or replacement of NASA real property, on a reimbursable basis within the Administration. Amounts in the fund are available without regard to fiscal year limitation. The capital of the fund consists of amounts appropriated to the fund; the reasonable value of stocks of supplies, equipment, and other assets and inventories on order that the Administrator transfers to the fund, less the related liabilities and unpaid obligations; amounts received from the sale of exchange of property; and payments received for loss or damage to property of the fund. The fund shall be reimbursed, in advance, for supplies and services at rates that will approximate the expenses of operation, such as the accrual of annual leave, depreciation of plant, property and equipment, and overhead.

Appointment of Commissioned Officer as Deputy Administrator
Pub. L. 107–117, div. B, § 307, Jan. 10, 2002, 115 Stat. 2301
Codified at 42 U.S.C. § 2472 note

SEC. 2472
During fiscal year 2002 the President, acting by and with the consent of the Senate, is authorized to appoint a commissioned officer of the Armed Forces, in active status, to the Office of Deputy Administrator of the National Aeronautics and Space Administration notwithstanding section 202(b) of the National Aeronautics and Space Act of 1958 (42 U.S.C. 2472 (b)). If so appointed, the provisions of section 403 (c)(3), (4), and (5) of title 50, United States Code, shall be applicable while the commissioned officer serves as Deputy Administrator in the same manner and extent as if the officer was serving in a position specified in section 403 (c) of title 50, United States Code, except that the officer's military pay and allowances shall be reimbursed from funds available to the National Aeronautics and Space Administration.

Notice of Reprogramming or Reorganization
Pub. L. 106–391, title III, § 311, Oct. 30, 2000, 114 Stat. 1594
Codified at 42 U.S.C. § 2473 note

SEC. 2473

(a) Notice of Reprogramming.—If any funds authorized by this Act [see Tables for classification] are subject to a reprogramming action that requires notice to be provided to the Appropriations Committees of the House of Representatives and the Senate, notice of such action shall concurrently be provided to the Committee on Science of the House of Representatives and the Committee on Commerce, Science, and Transportation of the Senate

(b) Notice of Reorganization.—The Administrator [of the National Aeronautics and Space Administration] shall provide notice to the Committees on Science and Appropriations of the House of Representatives, and the Committees on Commerce, Science, and Transportation and Appropriations of the Senate, not later than 30 days before any major reorganization of any program, project, or activity of the National Aeronautics and Space Administration.

Purchase of American-Made Equipment and Products
Pub. L. 106–391, title III, § 319, Oct. 30, 2000, 114 Stat. 1597
Codified at 42 U.S.C. § 2473 note

SEC. 2473

(a) Purchase of American-Made Equipment and Products.—In the case of any equipment or products that may be authorized to be purchased with financial assistance provided under this Act [see Tables for classification], it is the sense of the Congress that entities receiving such assistance should, in expending the assistance, purchase only American-made equipment and products.

(b) Notice to Recipients of Assistance.—In providing financial assistance under this Act, the Administrator [of the National Aeronautics and Space Administration] shall provide to each recipient of the assistance a notice describing the statement made in subsection (a) by the Congress.

Enhancement of Science and Mathematics Programs
Pub. L. 106–391, title III, § 321, Oct. 30, 2000, 114 Stat. 1597
Codified at 42 U.S.C. § 2473 note

SEC. 2473

(a) Definitions.—In this section:

(1) Educationally useful federal equipment.—The term 'educationally useful Federal equipment' means computers and related peripheral tools and research equipment that is appropriate for use in schools.

(2) School.—The term 'school' means a public or private educational institution that serves any of the grades of kindergarten through grade 12.

(b) Sense of the Congress

(1) In general.—It is the sense of the Congress that the Administrator [of the National Aeronautics and Space Administration] should, to the greatest extent practicable and in a manner consistent with applicable Federal law (including Executive Order No. 12999 [40 U.S.C. 549 note]), donate educationally useful Federal equipment to schools in order to enhance the science and mathematics programs of those schools.

(2) Reports.—Not later than 1 year after the date of the enactment of this Act [Oct. 30, 2000], and annually thereafter, the Administrator shall prepare and submit to Congress a report describing any donations of educationally useful Federal equipment to schools made during the period covered by the report.

NASA Flexibility Act of 2004
Pub. L. 108-201, § 2 (b), 118 Stat. 461, Feb. 24, 2004
Codified at 5 U.S.C. § 101 note, amended at 42 U.S.C. § 2473

This Act [adding Chapter 98 of Title 5 and amending 42 U.S.C. § 2473 and the part analysis preceding 5 U.S.C. § 2101] may be cited as the 'NASA Flexibility Act of 2004'.

Effective date. The amendment made by this section shall take effect on the first day of the first pay period beginning on or after the date of enactment of this of this [*sic*] Act.

APPENDIX C

A Half Century of NASA Spending 1959–2010: NASA Outlays in Relation to Total US Federal Government Outlays and to GDP

Year	Total US Federal Outlays in Current Dollars (*millions $*)	NASA Outlays in Current Dollars (*millions $*)	NASA Outlays as Share of Total US Federal Outlays (%)	NASA Outlays in Constant 2010 Dollars (*millions $*)	US GDP in Current Dollars (*billions $*)	NASA Outlays as Share of US GDP (%)
1959	92,098	146	0.16	871	506.6	0.03
1960	92,191	401	0.43	2370	526.4	0.08
1961	97,723	744	0.76	4340	544.8	0.14
1962	106,821	1257	1.18	7240	585.7	0.21
1963	111,316	2552	2.29	14,500	617.8	0.41
1964	118,528	4171	3.52	23,400	663.6	0.63
1965	118,228	5092	4.31	28,100	719.1	0.71
1966	134,532	5933	4.41	31,800	787.7	0.75
1967	157,464	5425	3.45	28,200	832.4	0.65
1968	178,134	4722	2.65	23,500	909.8	0.52
1969	183,640	4251	2.31	20,200	984.4	0.43
1970	195,649	3752	1.92	16,900	1,038.3	0.36
1971	210,172	3382	1.61	14,500	1,126.8	0.30
1972	230,681	3423	1.48	14,100	1,237.9	0.28
1973	245,707	3312	1.35	12,900	1,382.3	0.24
1974	269,359	3255	1.21	11,700	1,499.5	0.22
1975	332,332	3269	0.98	10,700	1,637.7	0.20

Sources: Office of Management and Budget Historical Tables 1.1 (for federal government outlays) and 4.1 (for NASA outlays 1962–2010), as of April 2011; *NASA Historical Data Book 1958–1968: Volume I, NASA Resources* (for NASA outlays 1959–1961); Bureau of Economic Analysis (for GDP data).

Year	Total US Federal Outlays in Current Dollars (*millions $*)	NASA Outlays in Current Dollars (*millions $*)	NASA Outlays as Share of Total US Federal Outlays (%)	NASA Outlays in Constant 2010 Dollars (*millions $*)	US GDP in Current Dollars (*billions $*)	NASA Outlays as Share of US GDP (%)
1976	371,792	3671	0.99	11,400	1,824.6	0.20
1977	409,218	4002	0.98	11,600	2,030.1	0.20
1978	458,746	4164	0.91	11,300	2,293.8	0.18
1979	504,028	4380	0.87	11,000	2,562.2	0.17
1980	590,941	4959	0.84	11,400	2,788.1	0.18
1981	678,241	5537	0.82	11,600	3,126.8	0.18
1982	745,743	6155	0.83	12,200	3,253.2	0.19
1983	808,364	6853	0.85	13,100	3,534.6	0.19
1984	851,805	7055	0.83	13,000	3,930.9	0.18
1985	946,344	7251	0.77	12,900	4,217.5	0.17
1986	990,382	7403	0.75	12,900	4,460.1	0.17
1987	1,004,017	7591	0.76	12,900	4,736.4	0.16
1988	1,064,416	9092	0.85	14,900	5,100.4	0.18
1989	1,143,744	11,036	0.96	17,400	5,482.1	0.20
1990	1,252,994	12,429	0.99	19,000	5,800.5	0.21
1991	1,324,226	13,878	1.05	20,500	5,992.1	0.23
1992	1,381,529	13,961	1.01	20,200	6,342.3	0.22
1993	1,409,386	14,305	1.01	20,200	6,667.4	0.21
1994	1,461,753	13,694	0.94	19,000	7,085.2	0.19
1995	1,515,742	13,378	0.88	18,200	7,414.7	0.18
1996	1,560,484	13,881	0.89	18,500	7,838.5	0.18
1997	1,601,116	14,360	0.90	18,800	8,332.4	0.17
1998	1,652,458	14,194	0.86	18,400	8,793.5	0.16
1999	1,701,842	13,636	0.80	17,400	9,353.5	0.15
2000	1,788,950	13,428	0.75	16,800	9,951.5	0.13
2001	1,862,846	14,092	0.76	17,200	10,286.2	0.14
2002	2,010,894	14,405	0.72	17,300	10,642.3	0.14
2003	2,159,899	14,610	0.68	17,200	11,142.1	0.13
2004	2,292,841	15,152	0.66	17,300	11,867.8	0.13
2005	2,471,957	15,602	0.63	17,300	12,638.4	0.12
2006	2,655,050	15,125	0.57	16,200	13,398.9	0.11
2007	2,728,686	15,861	0.58	16,500	14,061.8	0.11
2008	2,982,544	17,833	0.60	18,200	14,369.1	0.12
2009	3,517,677	19,168	0.54	19,400	14,119.0	0.14
2010	3,456,213	18,906	0.55	18,900	14,660.4	0.13

APPENDIX D

NASA Spending 1959–2010 (*millions $*)

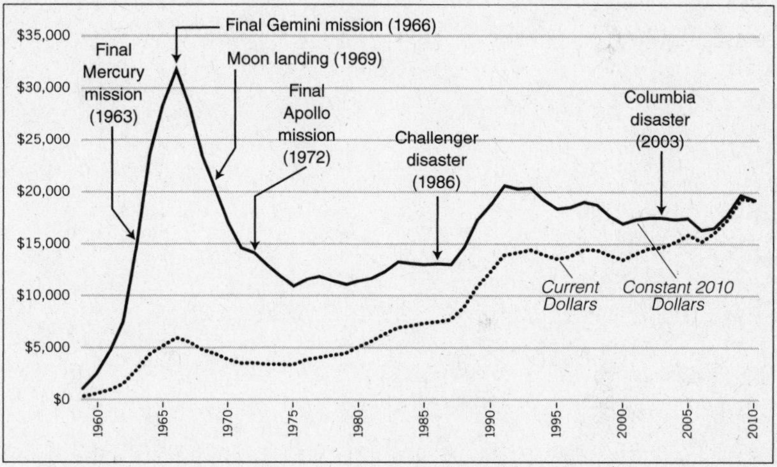

Sources: Office of Management and Budget Historical Tables 1.1 (for federal government outlays) and 4.1 (for NASA outlays 1962–2010), as of April 2011; *NASA Historical Data Book 1958–1968: Volume I, NASA Resources* (for NASA outlays 1959–1961).

APPENDIX E

NASA Spending as a Percentage of US Federal Government Spending and of US GDP 1959–2010

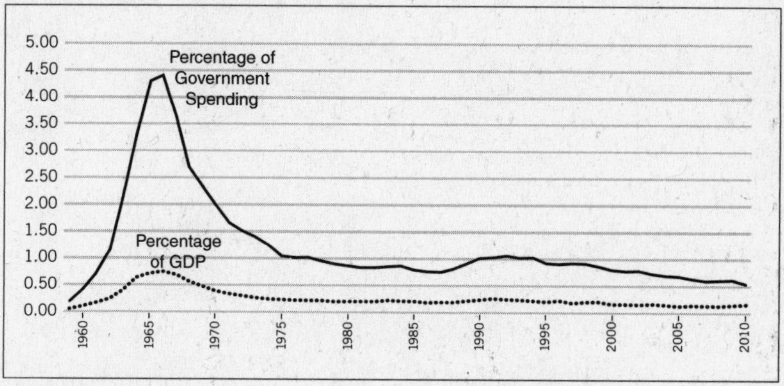

Sources: Office of Management and Budget Historical Tables 1.1 (for federal government outlays) and 4.1 (for NASA outlays 1962–2010), as of April 2011; *NASA Historical Data Book 1958–1968: Volume I, NASA Resources* (for NASA outlays 1959–1961); Bureau of Economic Analysis (for GDP data).

APPENDIX F

Space Budgets: US Government Agencies 2013

Agency	Budget	Source
Department of Defense (DoD)	$21.717 billion	DoD & Futron
National Aeronautics and Space Administration (NASA)	$16.865	NASA
National Oceanic and Atmospheric Administration (NOAA)	$1.886	NOAA
Department of Energy (DOE)	$0.060	DOE
Federal Aviation Administration (FAA)	$0.016	DOT
National Science Foundation (NSF)	$0.565	NSF
Federal Communications Commission (FCC)	$0.010	Estimate
Department of State	$0.003	DoS
United States Geological Survey (USGS)	$0.135	Department of Interior
Total	**$41.257 billion**	

Source: *The Space Report 2014*, © The Space Foundation, used with permission.

APPENDIX G

The Global Space Economy in 2013

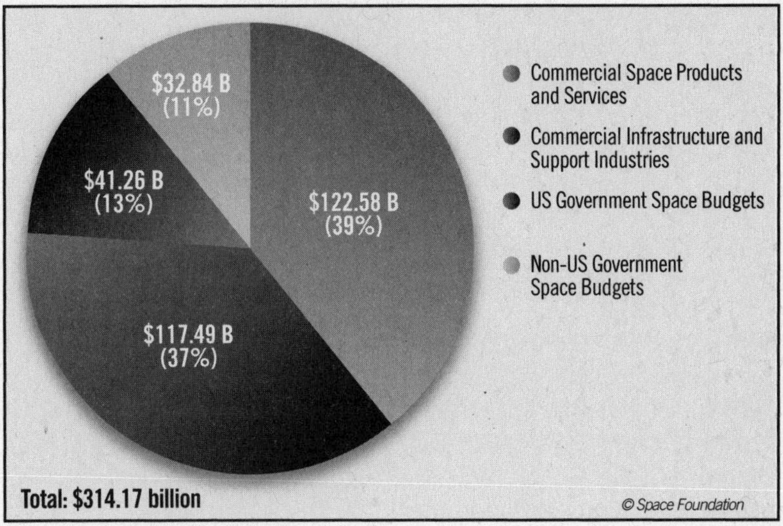

Source: *The Space Report 2014*, © The Space Foundation, used with permission.

APPENDIX H

Government Space Budgets 2013

Country/Agency	Budget (US $)	Source	Description
United States	$41.257 billion	[*see Appendix F*]	Fiscal Year 2013 Enacted
European Space Agency	$5.571 billion	European Space Agency	Calendar Year 2013 Appropriation
European Union*	$0.295 billion	European Commission	Calendar Year 2013 Appropriation
EUMETSAT*	$0.262	EUMETSAT	Calendar Year 2013 Spending
Brazil	$0.157 billion	Government of Brazil	Calendar Year 2013 Authorization
Canada*	$0.431 billion	Government of Canada	Fiscal Year 2013/2014 Appropriation
China	$3.468 billion	Estimate	Calendar Year 2013 Estimated Spending
France*	$0.966 billion	Government of France	Calendar Year 2013 Appropriation
Germany*	$0.786 billion	Government of Germany	Calendar Year 2013 Appropriation
India	$1.144 billion	Government of India	Fiscal Year 2013/2014 Allocation
Israel	$0.049 billion	*The Marker*	Calendar Year 2013 Estimated Spending
Italy*	$0.307 billion	Government of Italy	Calendar Year 2013 Planned Spending

Source: *The Space Report 2014*, © The Space Foundation, used with permission.

Country/Agency	Budget (US $)	Source	Description
Japan	$2.565 billion	Government of Japan	Fiscal Year 2013/2014 Appropriation
Russia	$5.482 billion	RIA Novosti	Calendar Year 2013 Planned Spending
South Korea	$0.304 billion	Korea Aerospace Research Institute (KARI)	Calendar Year 2013 Planned Spending
Spain*	$0.028 billion	Government of Spain	Calendar Year 2013 Appropriation
United Kingdom*	$0.087 billion	Estimated based on UK BIS Planning	Fiscal Year 2013/2014 Appropriation
Emerging Countries	$0.720 billion	[*multiple*]	
Non-US Military Space	$10.216 billion	Space Foundation estimate based on Euroconsult data	Calendar Year 2013 Estimated Spending
Total	**$74.095 billion**		

** Excludes ESA spending*

Note: Defense spending for all non-US countries is included in "non-US Military Space"

ACKNOWLEDGMENTS

Ann Rae Jonas transcribed most of the speeches contained herein, performing this task with a strong sense of not only what I said but, more important, what I meant. John M. Logsdon, a historian of space exploration without equal, provided valued information and insights. Richard W. Bulliet of Columbia University edited my very first essay on space exploration, "Paths to Discovery," which launched a subcareer of space commentary that continues to this day. Along the way, I've enjoyed conversations on our past, present, and future in space with astronauts Neil Armstrong, Buzz Aldrin, Tom Jones, Eileen Collins, and Kathy Sullivan; Congressman Robert Walker; author Andy Chaikin; scientists Steven Weinberg and Robert Lupton; and engineer Lou Friedman. I've further enjoyed conversations on national security with US Air Force generals Lester Lyles and John Douglass, US Navy commander Sue Hegg, and aerospace analyst Heidi Wood; and on NASA with space enthusiasts Lori Garver, Stephanie Schierholz, Elaine Walker, Elliott Pulham, and Bill Nye the Science Guy. I further recognize computer scientist Steve Napear for insightful conversations about the era of the great oceanic explorers and its correspondence with the era of space exploration. John Stockton offered useful comments and corrections to the first printing. Lastly, *Space Chronicles* would not exist without the support and enthusiasm for my work expressed by Avis Lang, longtime editor of my essays for *Natural History* magazine and editor of this volume.—NDT

• • •

Besides wanting to thank Neil Tyson for providing so many unexpected encounters with the cosmos, I am grateful for the literary and culinary assistance of Elliot Podwill; the graph-making skills of economist Anwar Shaikh; the perspective of Canadian space maven Surendra Parashar; the scrutiny of Norton Lang, Nivedita Majumdar, Fran Nesi, Julia Scully, and Eleanor Wachtel; and the troubleshooting of Elizabeth Stachow.—AL

INDEX

Page numbers in *italics* refer to illustrations.

BIOGRAPHICAL NOTES

About the Author

Neil deGrasse Tyson, an astrophysicist, was born and raised in New York City, where he was educated in the public schools clear through to his graduation from the Bronx High School of Science. He earned his BA in physics from Harvard and his PhD in astrophysics from Columbia. Tyson has served on two presidential commissions—one in 2001 on the future of the US aerospace industry, and a second in 2004 on the future of NASA—and on NASA's Advisory Council. Among his nine previous books are his memoir, *The Sky Is Not the Limit: Adventures of an Urban Astrophysicist*; the playful and informative *Death by Black Hole and Other Cosmic Quandaries*, which was a *New York Times* best seller; and *The Pluto Files: The Rise and Fall of America's Favorite Planet*. Tyson is the recipient of fourteen honorary doctorates and the NASA Distinguished Public Service Medal, the highest award given by the agency to a nongovernment civilian. His contributions to the public appreciation of the cosmos have been recognized by the International Astronomical Union in their official naming of asteroid 13123 Tyson. On the lighter side, he was voted "Sexiest Astrophysicist Alive" by *People* magazine in 2000. Tyson is the first occupant of the Hayden Planetarium's Frederick P. Rose directorship. He lives in New York City with his wife and two children.

About the Editor

Avis Lang is a writer, a freelance editor, and a lecturer in English at the City University of New York. She also collaborates with Neil deGrasse Tyson. From 2002 through 2007, as a senior editor at *Natural History* magazine, she oversaw Tyson's monthly column, "Universe." Originally trained as an art historian, Lang has written many essays on art and curated several large group exhibitions. Before moving to New York from Vancouver in 1983, she lectured for fifteen years at universities and art colleges across Canada.